Partial Differential Equations

Jae Kyong Cho

Professor
Department of Semiconductor Engineering
Gyeongsang National University, South Korea

Copyright © 2021 Jae Kyong Cho

All rights reserved.

CONTENTS

1. Wave Equations

1-1. One dimensional (1-D) Wave Equation

Modeling of a vibrating string -> 1-D wave equation
The Solution to the 1-D Wave Equation

1-2. Two dimensional (2-D) Wave Equation

Modeling of vibrating membrane -> 2-D wave equation
Modeling of vibrating rectangular membrane
-> 2-D wave equation in rectangular coordinate system

2. Heat Equations

2-1. 1-D Heat Equation

Modeling of heat conduction inside a long and thin rod
-> 1-D heat equation
The Solution to the 1-D Heat Equation

2-2. 2-D Heat Equation

Modeling of heat conduction inside membrane
-> 2-D heat equation
The heat equation in a steady state -> Laplace's equation

3. Laplace's equation

Gravity, Electrostatic Potential -> 3-D Laplace's Equation

4. The Solutions to PDEs by means of the Integral Transform

The solutions to PDEs by means of the Laplace Transform
The solution to PDE by means of the Fourier Transform

PATIAL DIFFERENTIAL EQUATIONSS

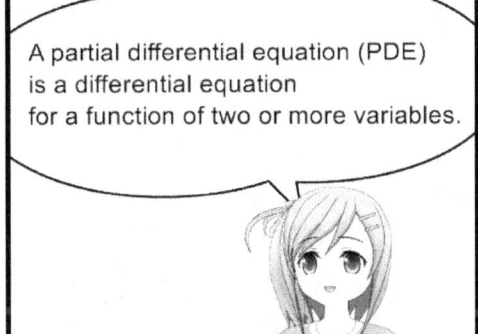

A partial differential equation (PDE) is a differential equation for a function of two or more variables.

Since our world is 4-D time-space, if we put in a function what's happening in the world, we get a function of many variables that has a time variable (t) and three spatial variables (x, y, z).

u(x, y, z, t)

What we are interested in is changes that occur in the world and each of the changes has many variables, and therefore, we should make clear which the change is about (which particular differential)?

$\frac{\partial u}{\partial t}, \frac{\partial u}{\partial x}$

So, we get an equation having partial derivatives, and the equation is called a partial differential equation (PDE, for short).

Such a partial differential equation has much more areas of application than the ordinary differential equation (ODE, for short, which is a differential equation for a function of one variable only).

For instance, the areas of application include fluid mechanics, solid mechanics, heat transfer, electromagnetism, quantum mechanics, etc.

Ah! And yet! The most important in PDE are

Wave Equation $\quad \frac{\partial^2 u}{\partial t^2} = \nabla^2 u$

Heat Equation $\quad \frac{\partial u}{\partial t} = \nabla^2 u$

Laplace Equation $\quad \nabla^2 u = 0$

PARTIAL DIFFERENTIAL EQUATIONS

PATIAL DIFFERENTIAL EQUATIONS

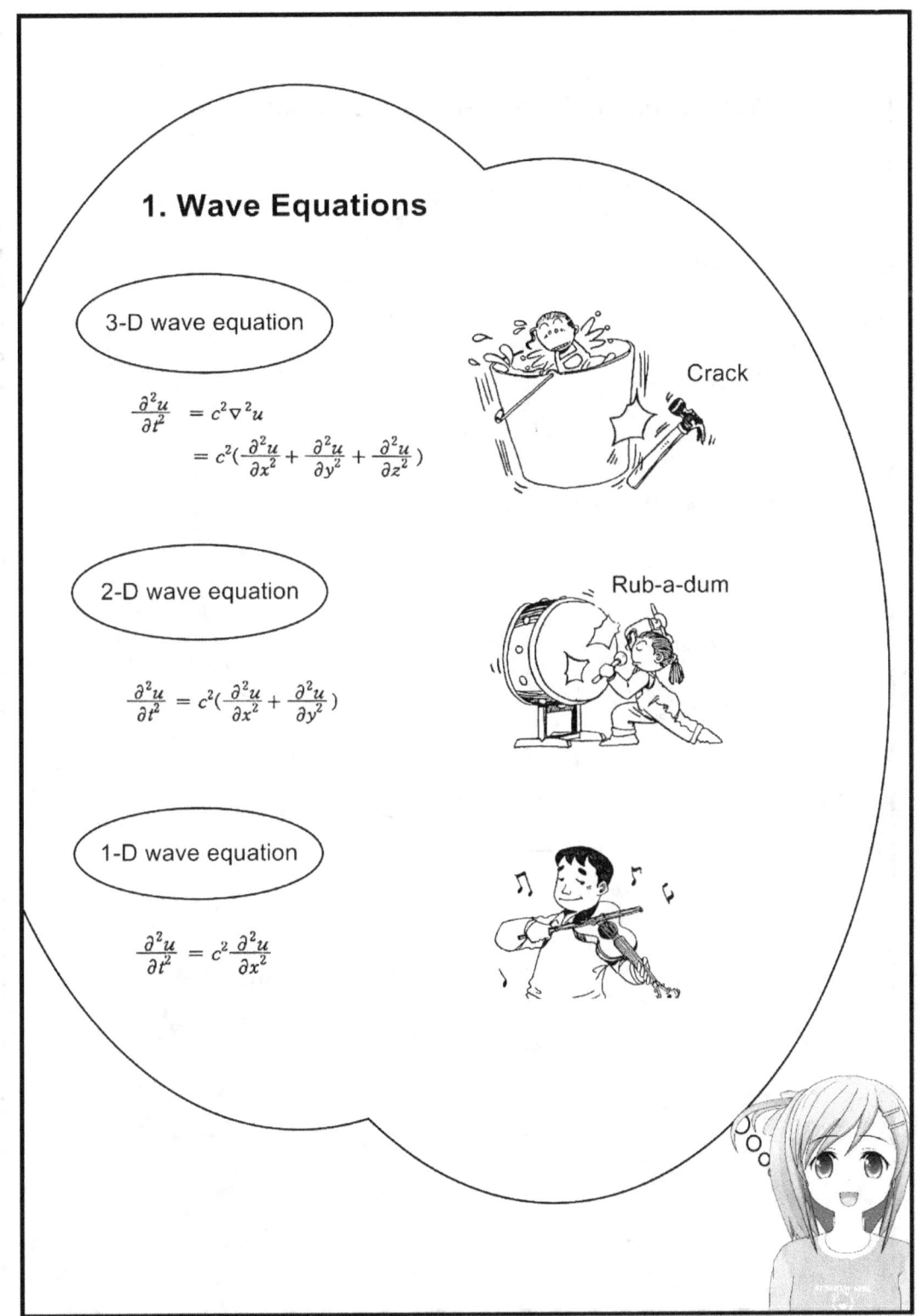

1. Wave Equations

3-D wave equation

$$\frac{\partial^2 u}{\partial t^2} = c^2 \nabla^2 u$$
$$= c^2 \left(\frac{\partial^2 u}{\partial x^2} + \frac{\partial^2 u}{\partial y^2} + \frac{\partial^2 u}{\partial z^2} \right)$$

Crack

2-D wave equation

$$\frac{\partial^2 u}{\partial t^2} = c^2 \left(\frac{\partial^2 u}{\partial x^2} + \frac{\partial^2 u}{\partial y^2} \right)$$

Rub-a-dum

1-D wave equation

$$\frac{\partial^2 u}{\partial t^2} = c^2 \frac{\partial^2 u}{\partial x^2}$$

PATIAL DIFFERENTIAL EQUATIONSS

1-1. One dimensional (1-D) Wave Equation

3-D wave equation

$$\frac{\partial^2 u}{\partial t^2} = c^2 \nabla^2 u$$

where

$$\nabla^2 u = \frac{\partial^2 u}{\partial x^2} + \frac{\partial^2 u}{\partial y^2} + \frac{\partial^2 u}{\partial z^2}$$

Since it is 1-D, we get

$$\nabla^2 u = \frac{\partial^2 u}{\partial x^2}$$

Consequently, the 1-D wave equation is

$$\frac{\partial^2 u}{\partial t^2} = c^2 \frac{\partial^2 u}{\partial x^2} \qquad \left(c^2 = \frac{T}{\rho} \right)$$

PATIAL DIFFERENTIAL EQUATIONS

Modeling of a vibrating string: 1-D wave equation

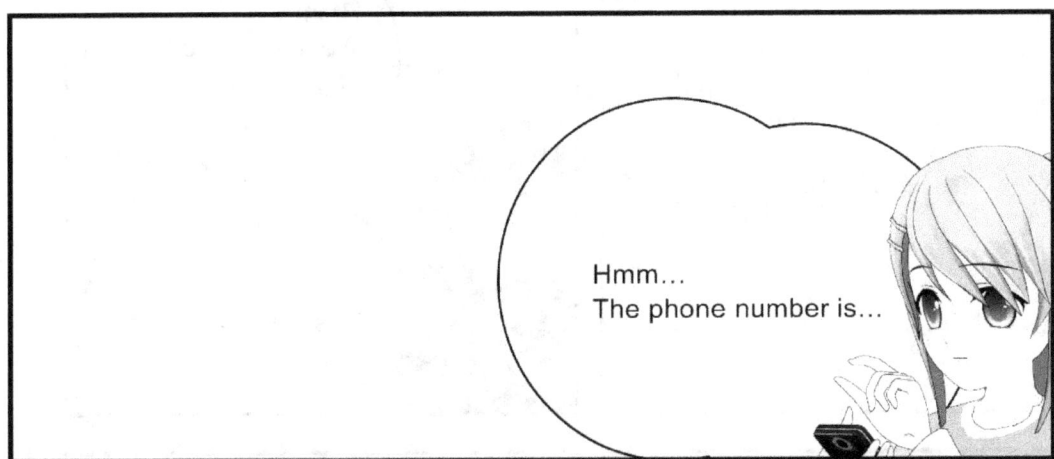

PARTIAL DIFFERENTIAL EQUATIONS

PATIAL DIFFERENTIAL EQUATIONS

Blowing it up, we get this.

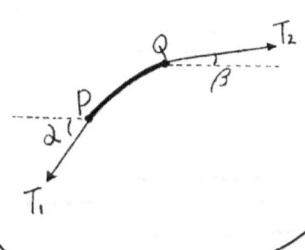

Let's consider the horizontal and vertical components of the tensions.

1) Horizontal components:
Since no vibration occurs in the horizontal direction, the horizontal components of the two tensions have to be the same. That is, Setting the above equal to be T, we get

$$T_2 \cos \beta = T_1 \cos \alpha = T \quad \cdots \cdots (1)$$

2) Vertical components:
Since vibrations occur in the vertical direction, the Newton's equation of motion applies.
All the forces

$$F = T_2 \sin \beta - T_1 \sin \alpha$$

This has to be the same as ma,
where mass m = $\rho \Delta x$
(rho is the density and delta x is the length)
and $a = \frac{\partial^2 u}{\partial t^2}$ (u is the displacement)
Therefore, we get

$$T_2 \sin \beta - T_1 \sin \alpha = \rho \Delta x \frac{\partial^2 u}{\partial t^2} \quad \cdots (2)$$

Division of the expression (2) by the expression (1) yields

$$\tan \beta - \tan \alpha = \frac{\rho \Delta x}{T} \frac{\partial^2 u}{\partial t^2}$$

Since tan alpha and tan beta are the slopes at the point x and the point x + Δx, they have to be the same as the derivatives at these points. Therefore, we get

$$\left(\frac{\partial u}{\partial x}\right)_x - \left(\frac{\partial u}{\partial x}\right)_{x+\Delta x}$$
$$= \frac{\rho \Delta x}{T} \frac{\partial^2 u}{\partial t^2}$$

PARTIAL DIFFERENTIAL EQUATIONS

Dividing both sides by delta x, we get

$$\frac{1}{\Delta x}\left[\left(\frac{\partial u}{\partial x}\right)_x - \left(\frac{\partial u}{\partial x}\right)_{x+\Delta x}\right] = \frac{\rho}{T}\frac{\partial^2 u}{\partial t^2}$$

If we send $\Delta x \to 0$, the left hand side will be

$$\frac{\partial^2 u}{\partial x^2}$$

Finally, we get

$$\boxed{\frac{\partial^2 u}{\partial t^2} = c^2 \frac{\partial^2 u}{\partial x^2}}$$

where $c^2 = \frac{T}{\rho}$

The reason that we set $\frac{T}{\rho}$ equal to c^2 is that we make clear that $\frac{T}{\rho}$ gets a positive value.

This is called 1-D wave equation.

We can put in the 1-D wave equation the vibrations of all the strings of string instruments such as the vibration of the guitar string.

PATIAL DIFFERENTIAL EQUATIONS

The Solution to the 1-D Wave Equation

If we want to see how the guitar string vibrates (how the displacement changes), we solve the 1-D wave equation and find the solution u(x, t), don't we?

Right! Now, the 1-D wave equation is a PDE. How do we solve it, then?

We can use the separation of variables method!

Separation of Variables Method

1. Suppose the solution is (a function of x only) X (a function of t only).
2. Put it into the PDE given and make two ordinary differential equations (ODE). (ODE with respect to x only and ODE with respect to t only)
3. Find the solutions to the two ODEs.
4. Take the product of the two solutions!

PARTIAL DIFFERENTIAL EQUATIONS

Um... Let's actually solve $\dfrac{\partial^2 u}{\partial t^2} = c^2 \dfrac{\partial^2 u}{\partial x^2}$

1. Taking apart the PDE into two ODEs

Assume that the solution of the PDE, u(x, t) is

$$u(x,\ t) = F(x)\,G(t)$$

Putting them into the PDE given, we get

$$F\ddot{G} = c^2 F'' G$$

Assorting them, we get

$$\frac{1}{c^2}\frac{\ddot{G}}{G} = \frac{F''}{F}$$

where the double dot indicates the 2nd derivative with respect to time t and the double dash indicates the 2nd derivative with respect to x.

PATIAL DIFFERENTIAL EQUATIONS

The left hand side is a function of t only, and the right hand side is a function of x only.
Um...
If these two sides have to be equal, they have to be constants.

Right!
Assuming the constant is k, we get this, don't we?

$$\frac{1}{c^2} \frac{\ddot{G}}{G} = \frac{F''}{F} = k$$

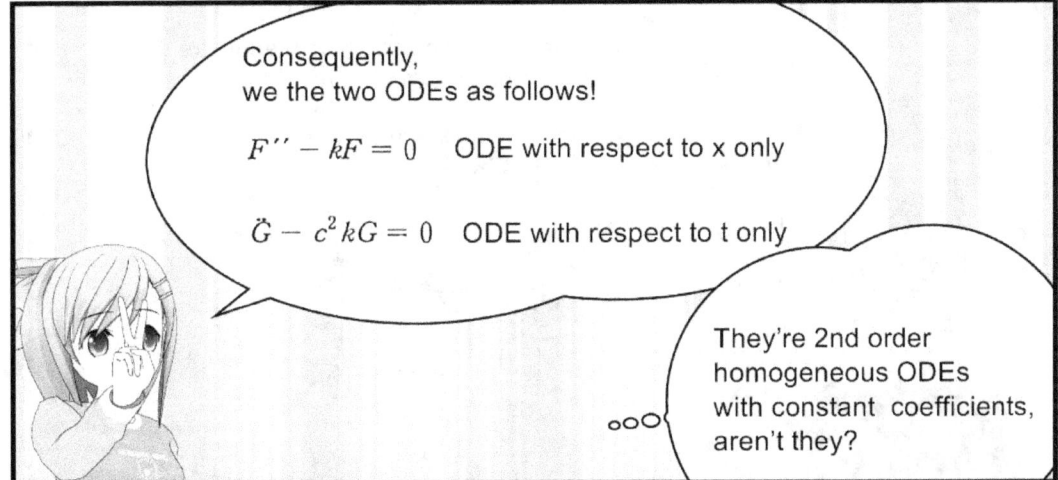

Consequently, we the two ODEs as follows!

$F'' - kF = 0$ ODE with respect to x only

$\ddot{G} - c^2 kG = 0$ ODE with respect to t only

They're 2nd order homogeneous ODEs with constant coefficients, aren't they?

PARTIAL DIFFERENTIAL EQUATIONS

PATIAL DIFFERENTIAL EQUATIONS

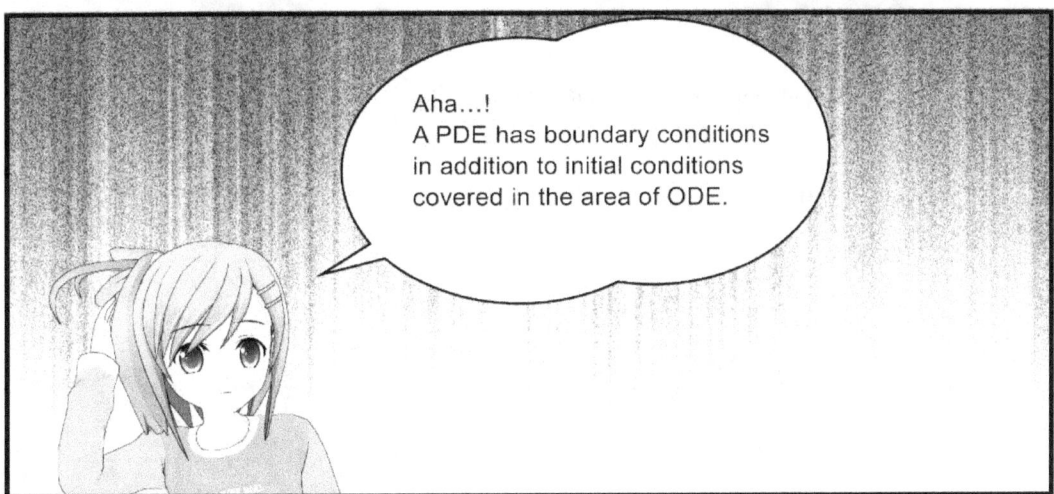

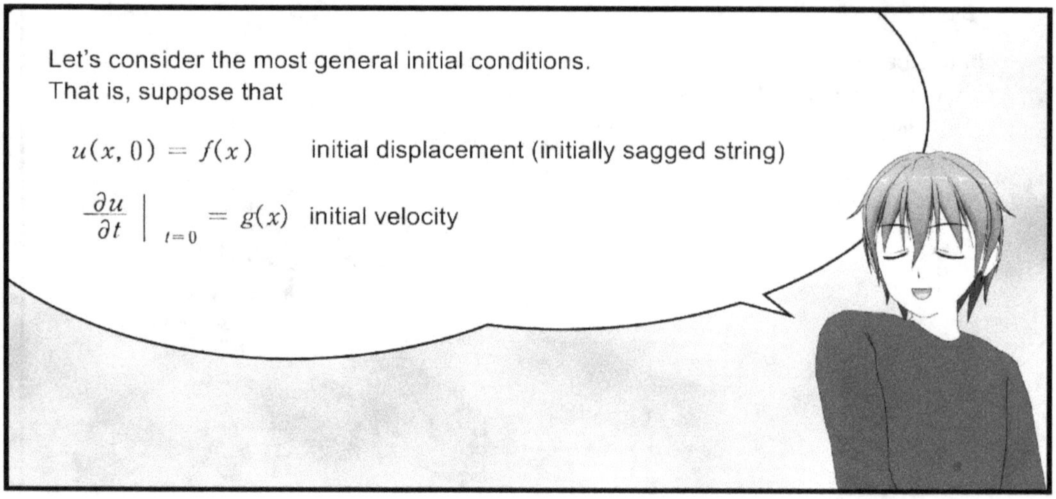

PARTIAL DIFFERENTIAL EQUATIONS

Let's consider the most general initial conditions. That is, suppose that

$u(x, 0) = f(x)$ initial displacement (initially sagged string)

$\left.\dfrac{\partial u}{\partial t}\right|_{t=0} = g(x)$ initial velocity

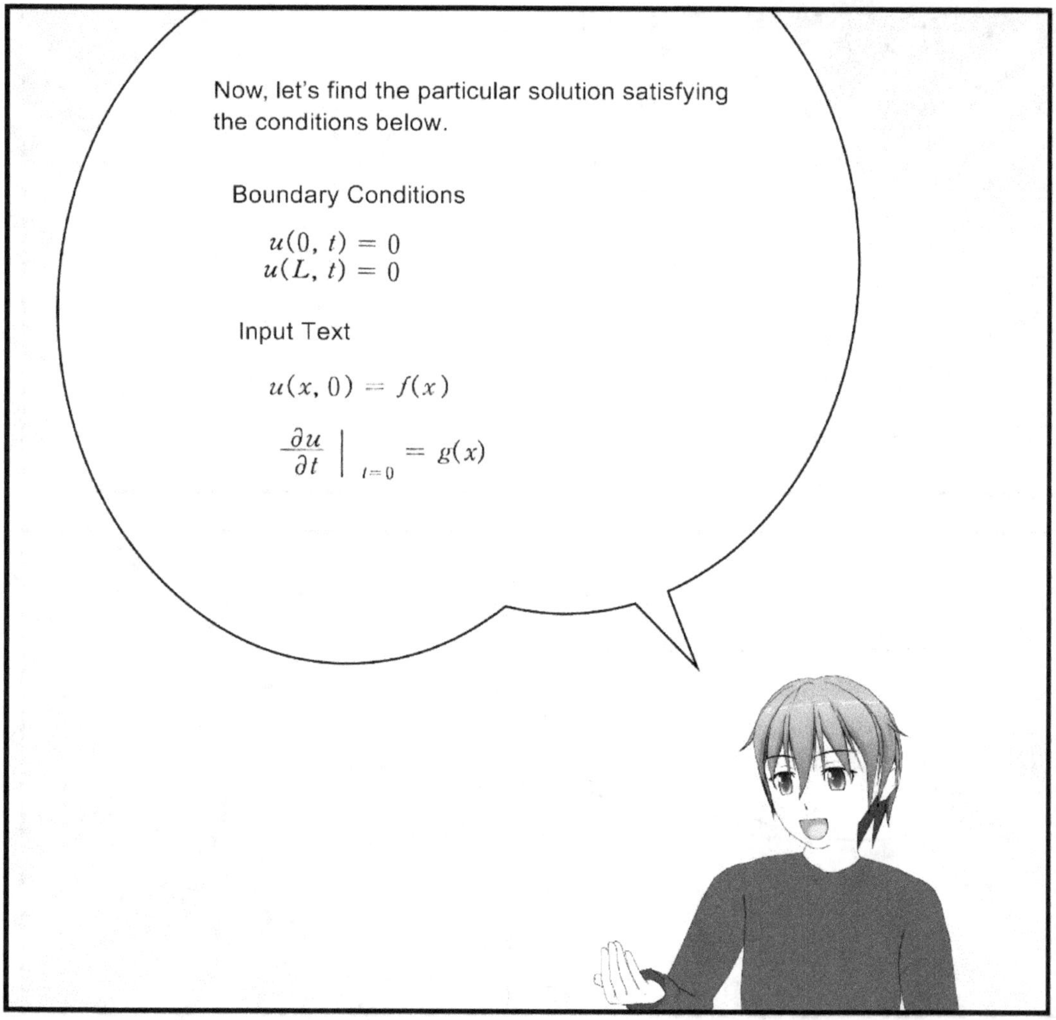

Now, let's find the particular solution satisfying the conditions below.

Boundary Conditions

$u(0, t) = 0$
$u(L, t) = 0$

Input Text

$u(x, 0) = f(x)$

$\left.\dfrac{\partial u}{\partial t}\right|_{t=0} = g(x)$

PATIAL DIFFERENTIAL EQUATIONS

2. Satisfying Boundary Conditions

Before we solve the two ODEs

$$F'' - kF = 0$$

$$\ddot{G} - c^2 kG = 0$$

found earlier and find $u(x, t) = F(x) G(t)$ that we want,

let's analyze the boundary conditions
$$u(0, t) = 0$$
$$u(L, t) = 0$$

Then, we can see
$$u(0, t) = F(0) G(t) = 0$$
$$u(L, t) = F(L) G(t) = 0$$

Now, from the first equation, it must be that F(0) = 0 or G(t) = 0!

17

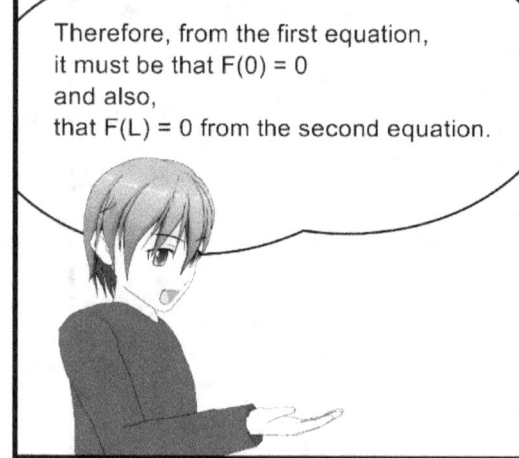

PATIAL DIFFERENTIAL EQUATIONS

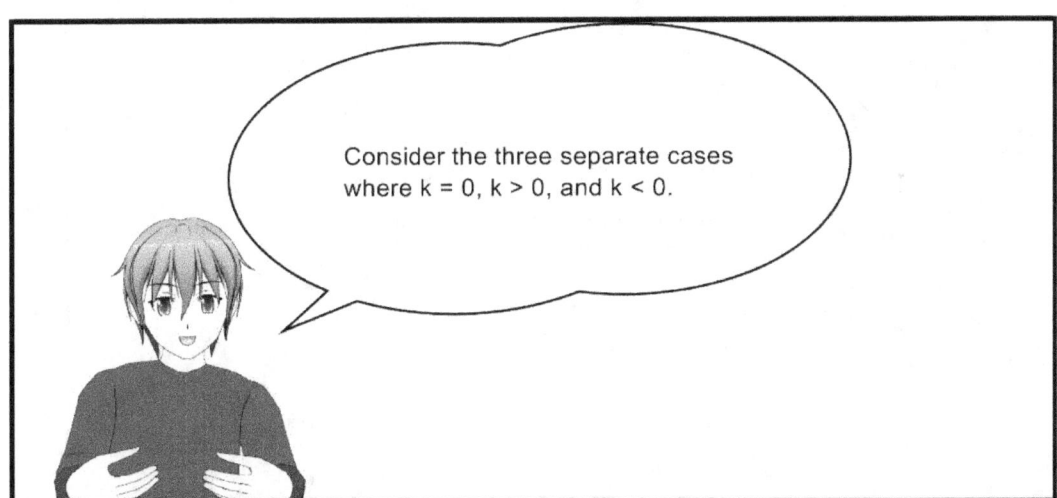

PARTIAL DIFFERENTIAL EQUATIONS

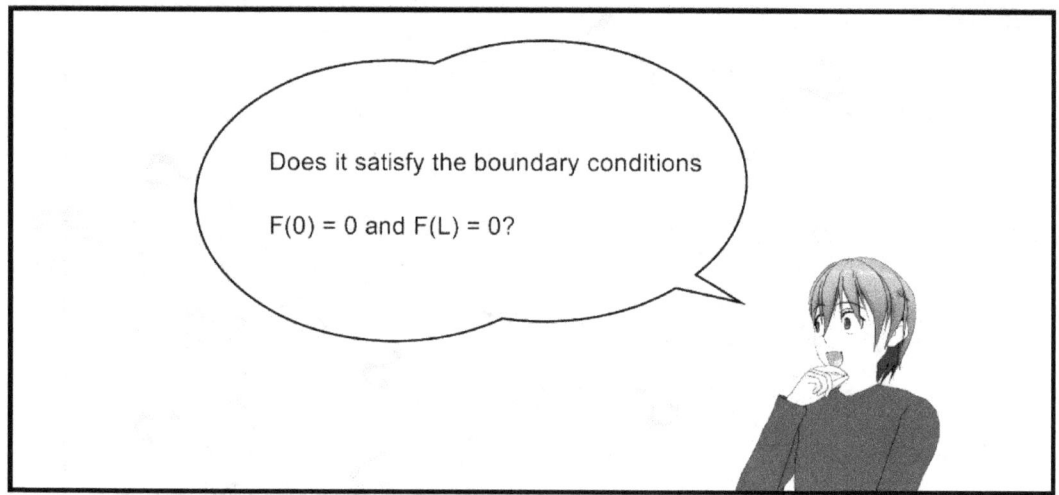

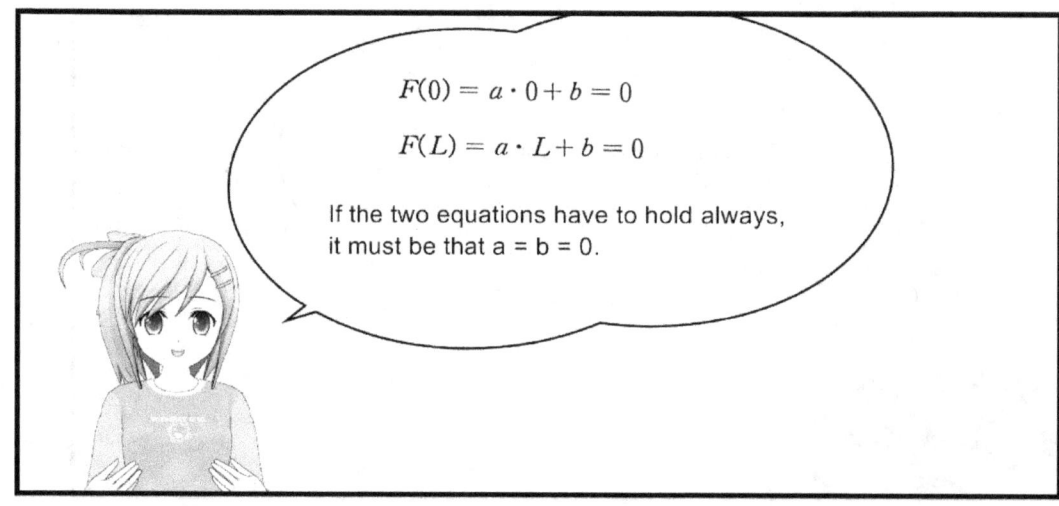

PATIAL DIFFERENTIAL EQUATIONS

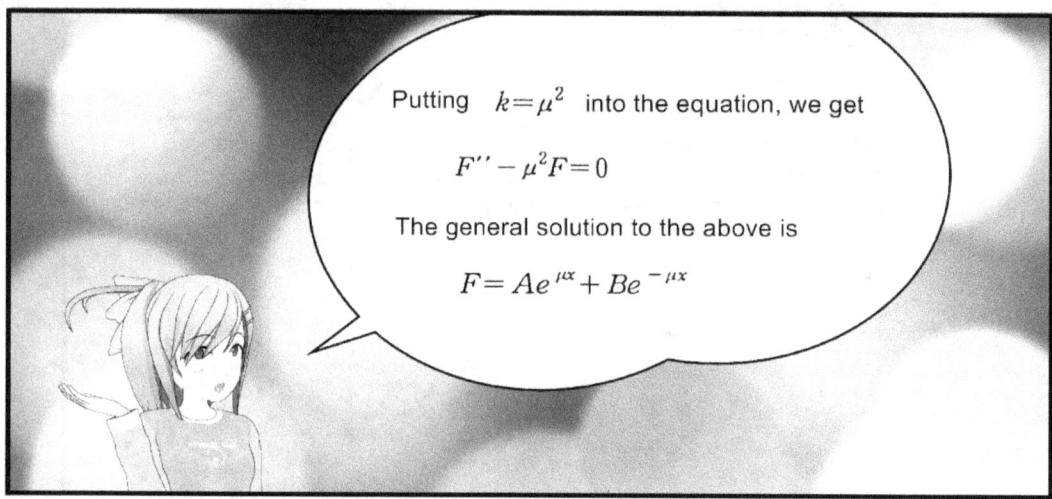
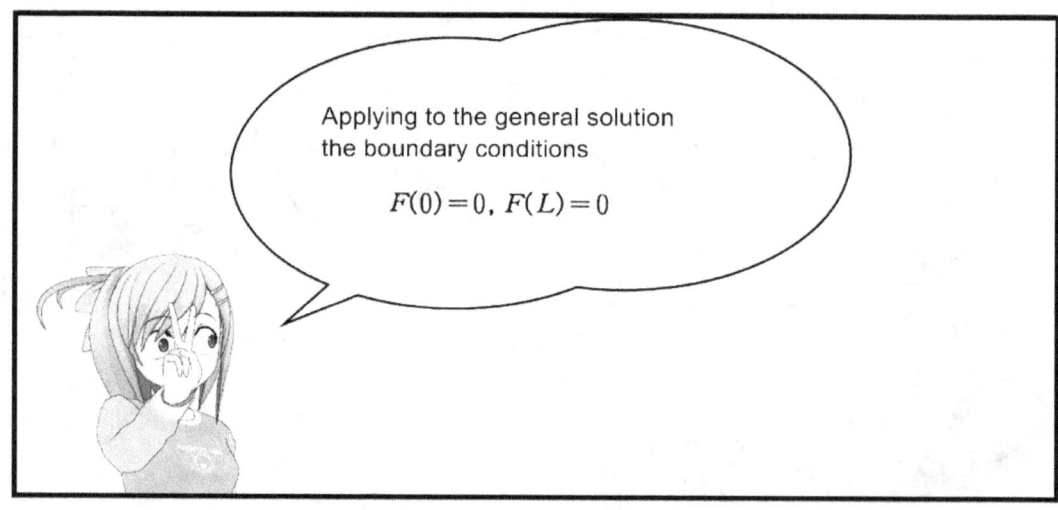

PATIAL DIFFERENTIAL EQUATIONS

$$F(0) = Ae^{\mu \cdot 0} + Be^{-\mu \cdot 0} = A + B = 0$$

$$F(L) = Ae^{\mu L} + Be^{-\mu L} = 0$$

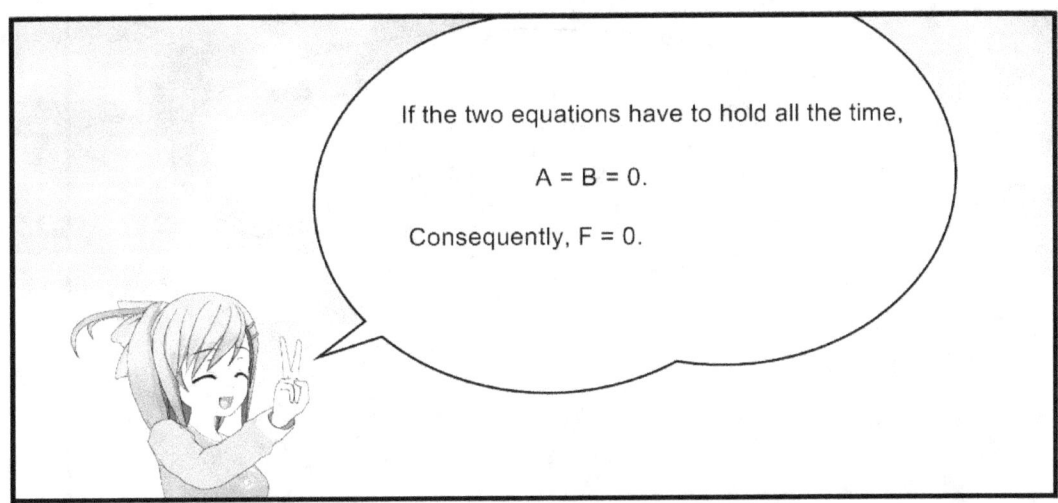

If the two equations have to hold all the time,

$A = B = 0$.

Consequently, $F = 0$.

Therefore, $u = 0$. Eh, likewise!
It doesn't... make sense....
(Nonsense)

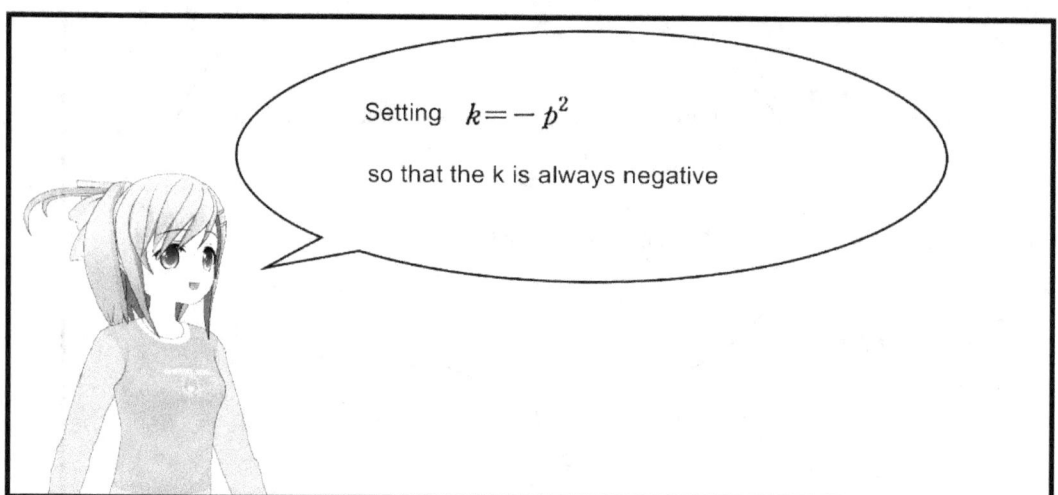

PATIAL DIFFERENTIAL EQUATIONS

If B = 0 in

$$F(L) = B\sin pL = 0$$

F=0. -> Nonsense!

Consequently, sin pL = 0. Therefore

$$pL = n\pi$$
$$p = \frac{n\pi}{L}$$

So, putting

$$A = 0,\ p = \frac{n\pi}{L}$$

into the general solution,

we get

$$F_n(x) = B\sin\frac{n\pi}{L}x\ (n=1,\ 2,\cdots)$$

which is the only one that works.

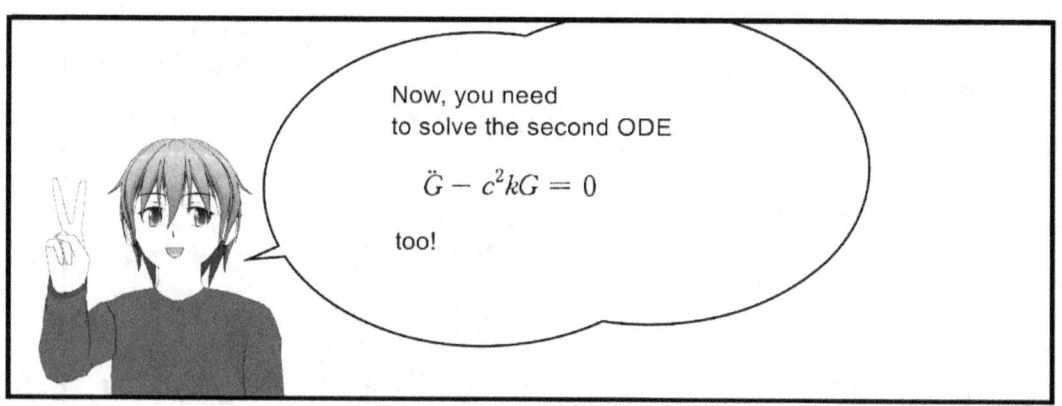

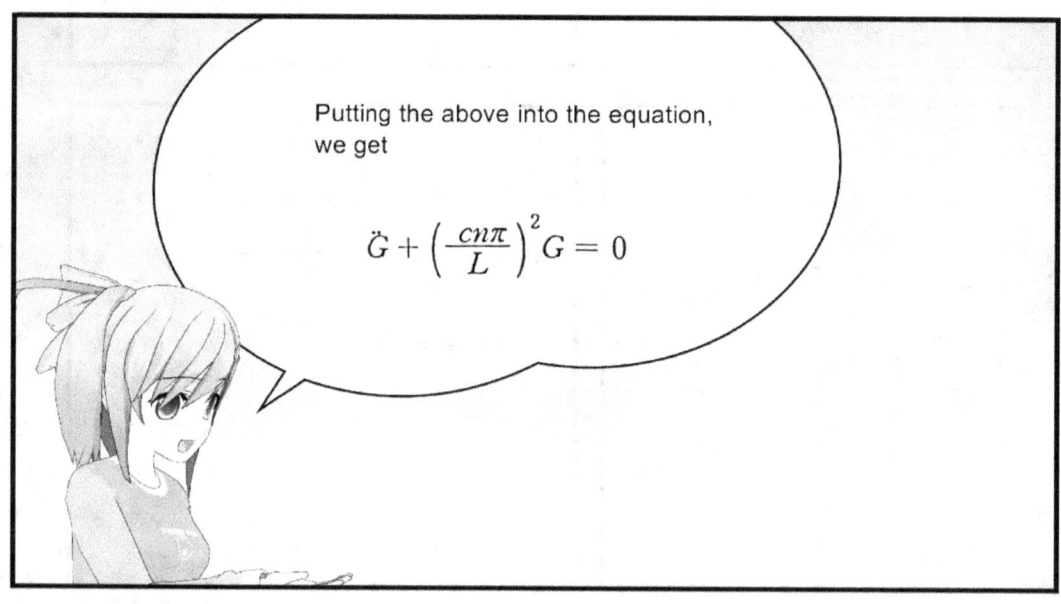

PATIAL DIFFERENTIAL EQUATIONS

Setting
$$\left(\frac{cn\pi}{L}\right)^2$$
equal to
$$\lambda_n^2$$
we get
$$\ddot{G} + \lambda_n^2 G = 0$$

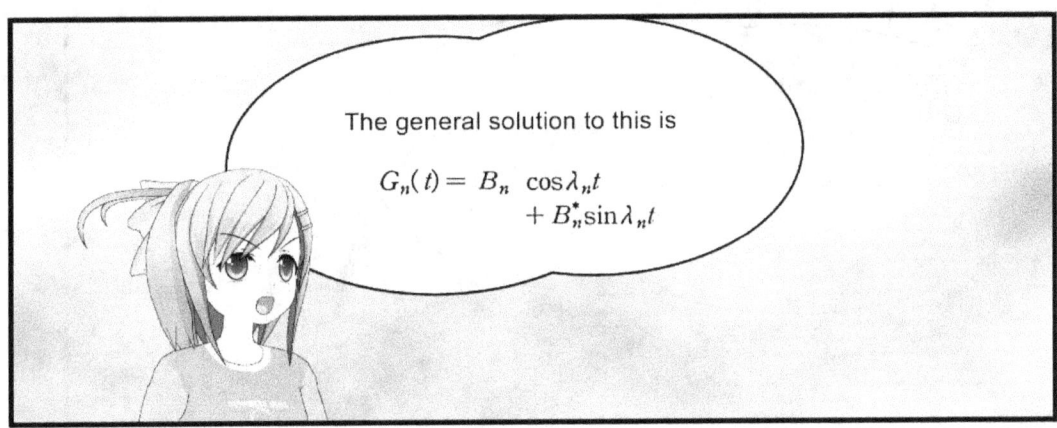

The general solution to this is
$$G_n(t) = B_n \cos\lambda_n t + B_n^* \sin\lambda_n t$$

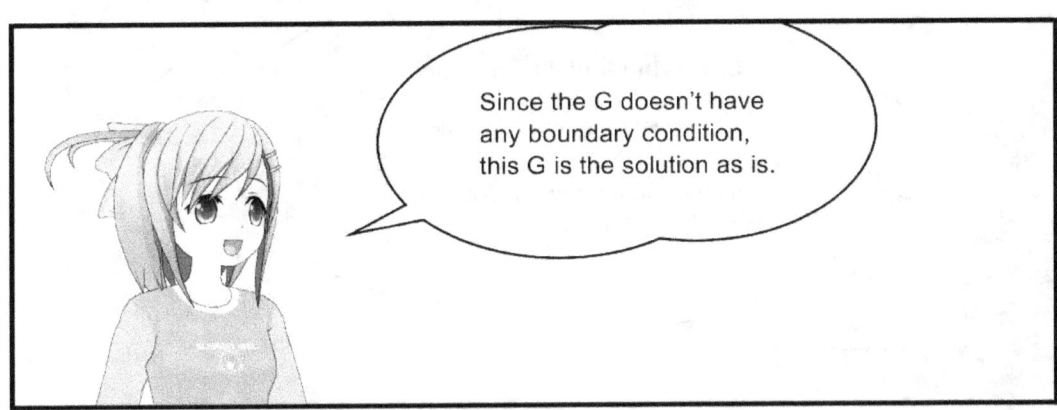

Since the G doesn't have any boundary condition, this G is the solution as is.

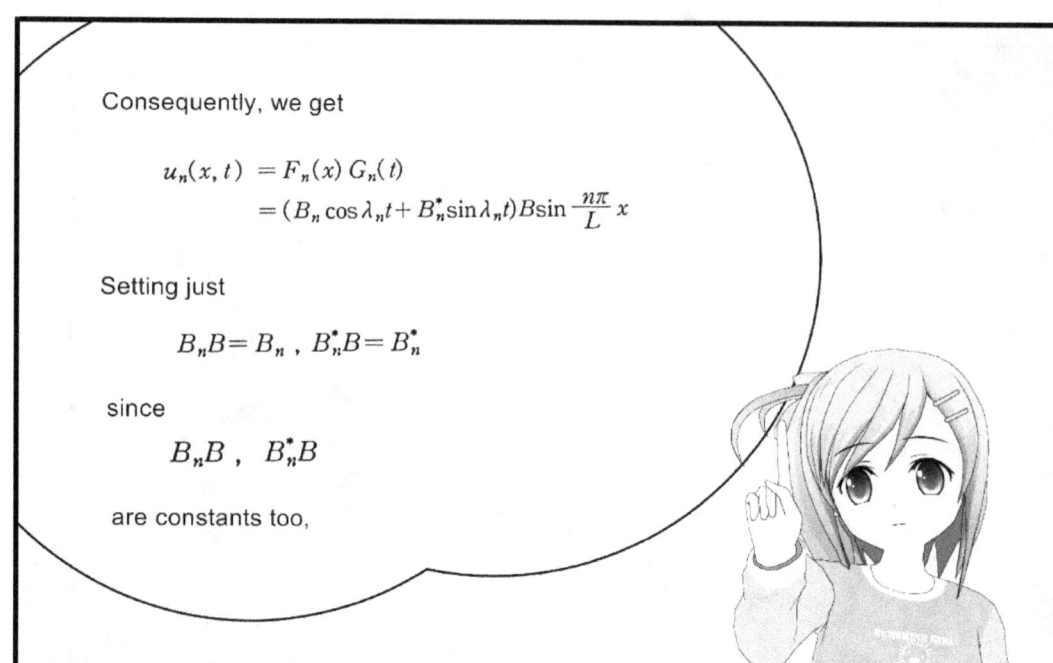
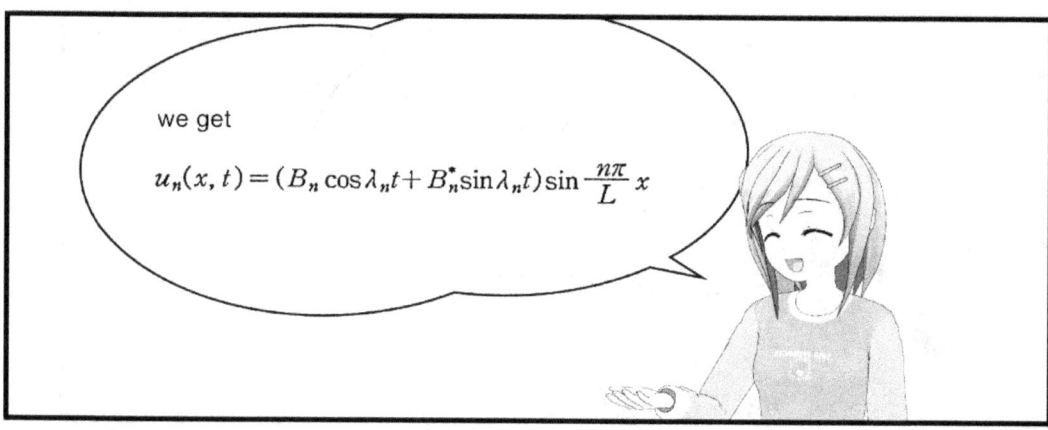
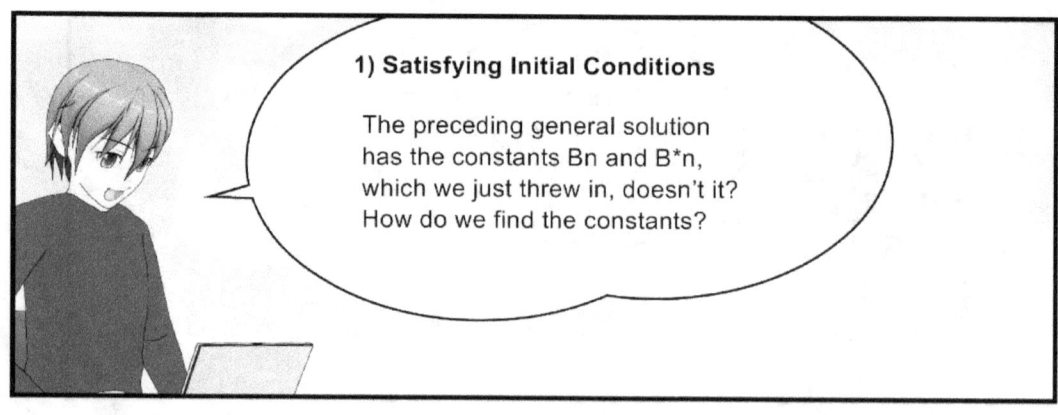

PATIAL DIFFERENTIAL EQUATIONS

PARTIAL DIFFERENTIAL EQUATIONS

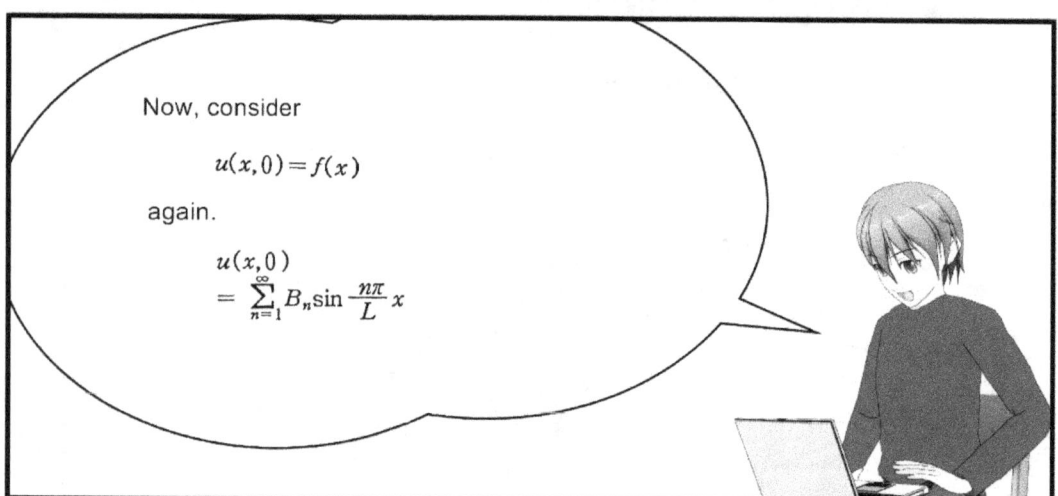

Since the 1st order wave equation is linear and homogeneous, the sum of all the solutions un(x, t) (n = 1, 2, ...) too can be a solution, can't it?
That is, putting the sum in u(x, t), we get

$$u(x,t) = \sum_{n=1}^{\infty} u_n(x,t)$$
$$= \sum_{n=1}^{\infty}(B_n\cos\lambda_n t + B_n^*\sin\lambda_n t)\sin\frac{n\pi}{L}x$$

Now, consider

$$u(x,0) = f(x)$$

again.

$$u(x,0) = \sum_{n=1}^{\infty} B_n \sin\frac{n\pi}{L}x$$

Then, the above can be equal to f(x) since it is in the form of the expansion of f(x) into the Fourier series.

That is,

$$u(x,0) = \sum_{n=1}^{\infty} B_n \sin\frac{n\pi}{L}x = f(x)$$

PATIAL DIFFERENTIAL EQUATIONS

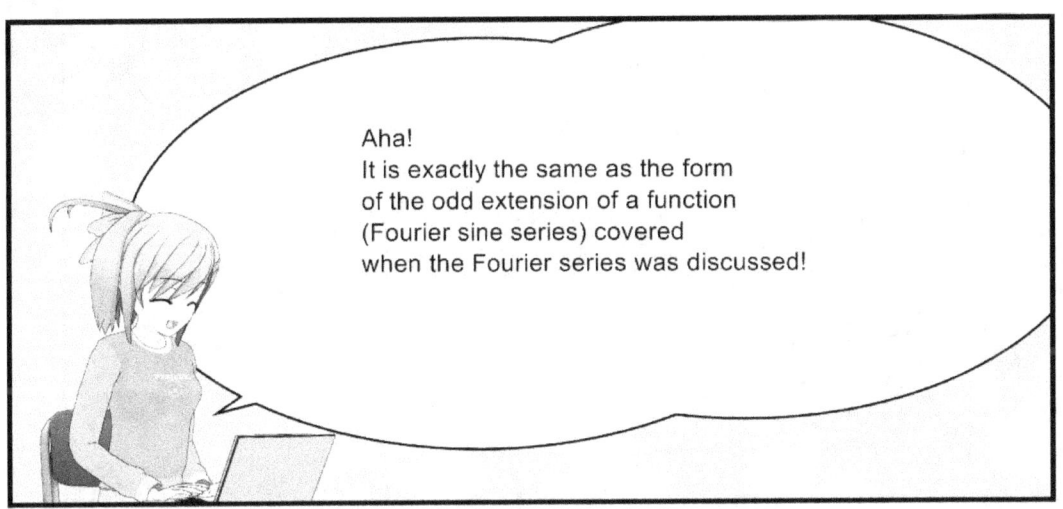

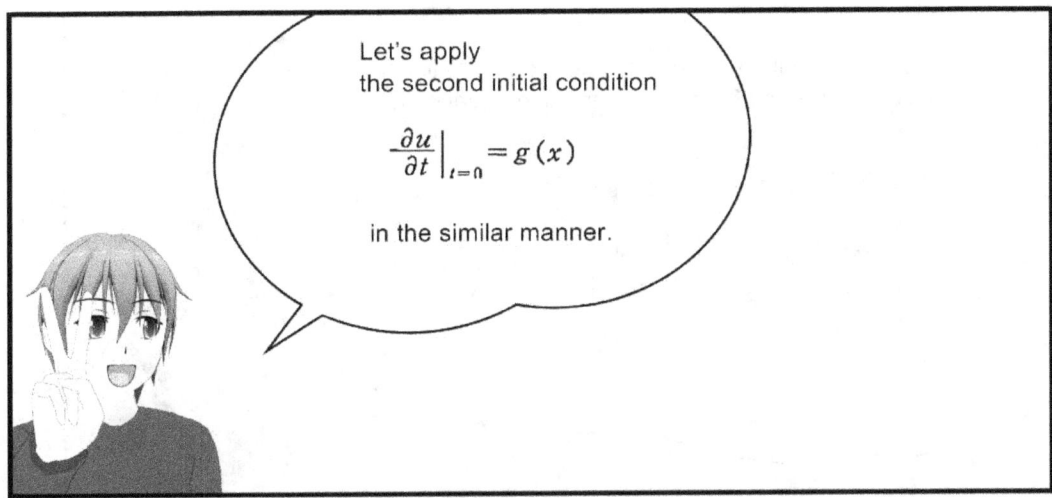

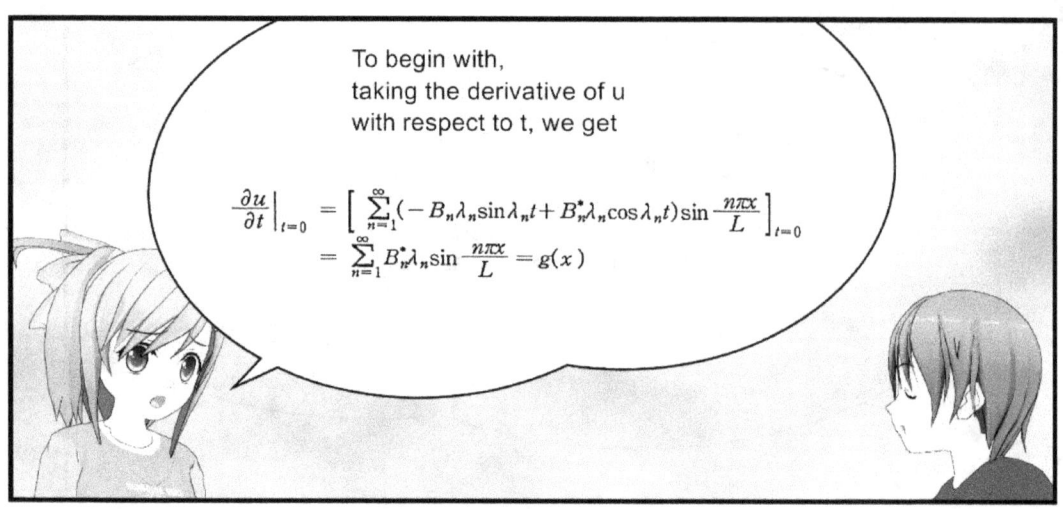

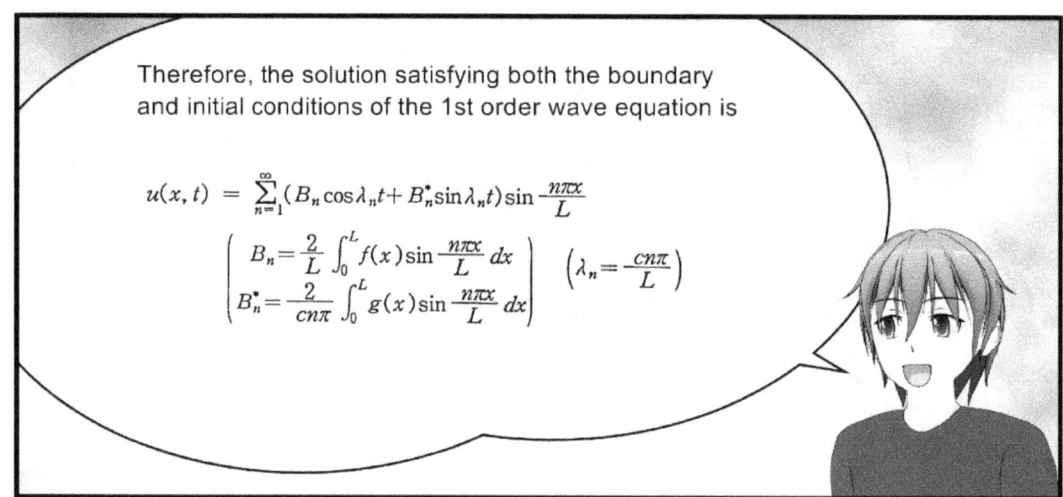

PATIAL DIFFERENTIAL EQUATIONS

So, if we are given the initial conditions, that is, the initial displacement f(x) and the initial velocity g(x), we can put them into the formula above and readily get the solution.

What will the solution be if the initial velocity g(x) = 0?

Since B*n = 0, we get

$$u(x,t) = \sum_{n=1}^{\infty} B_n \cos \lambda_n t \sin \frac{n\pi x}{L}$$

$$\left(\begin{array}{l} B_n = \frac{2}{L} \int_0^L f(x) \sin \frac{n\pi x}{L} \, dx \\ \lambda_n = \frac{cn\pi}{L} \end{array} \right)$$

We can convert the equation above by the trigonometric arithmetic theorem in such a way as follows.

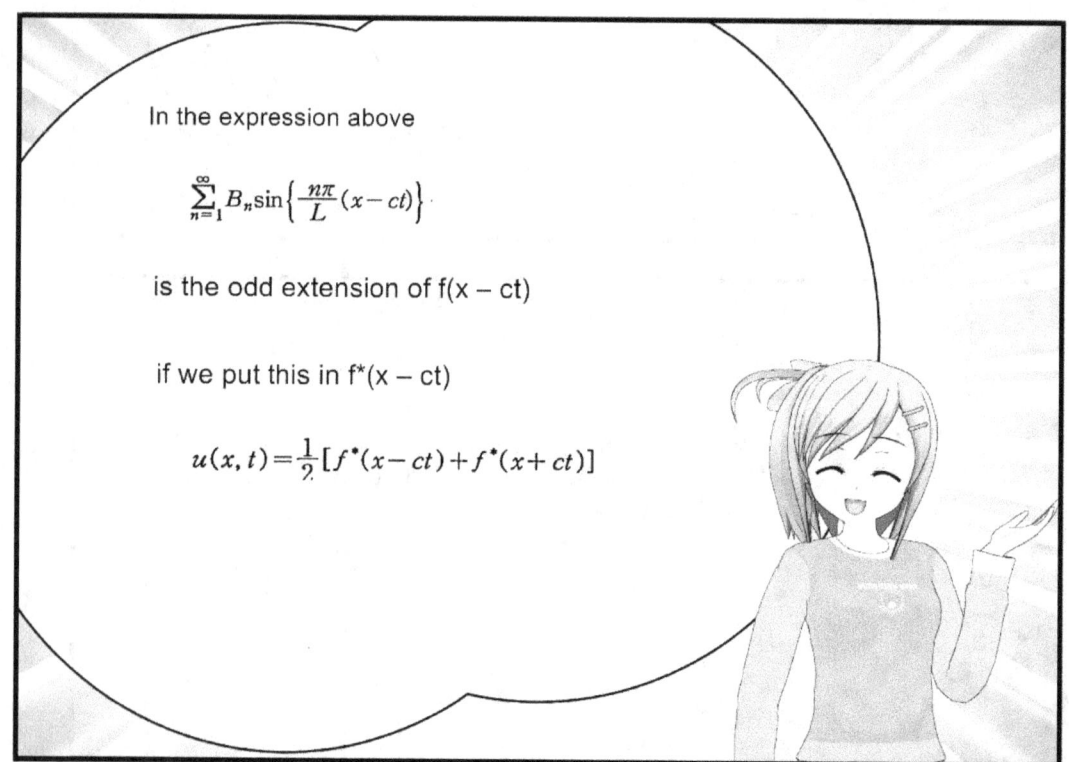

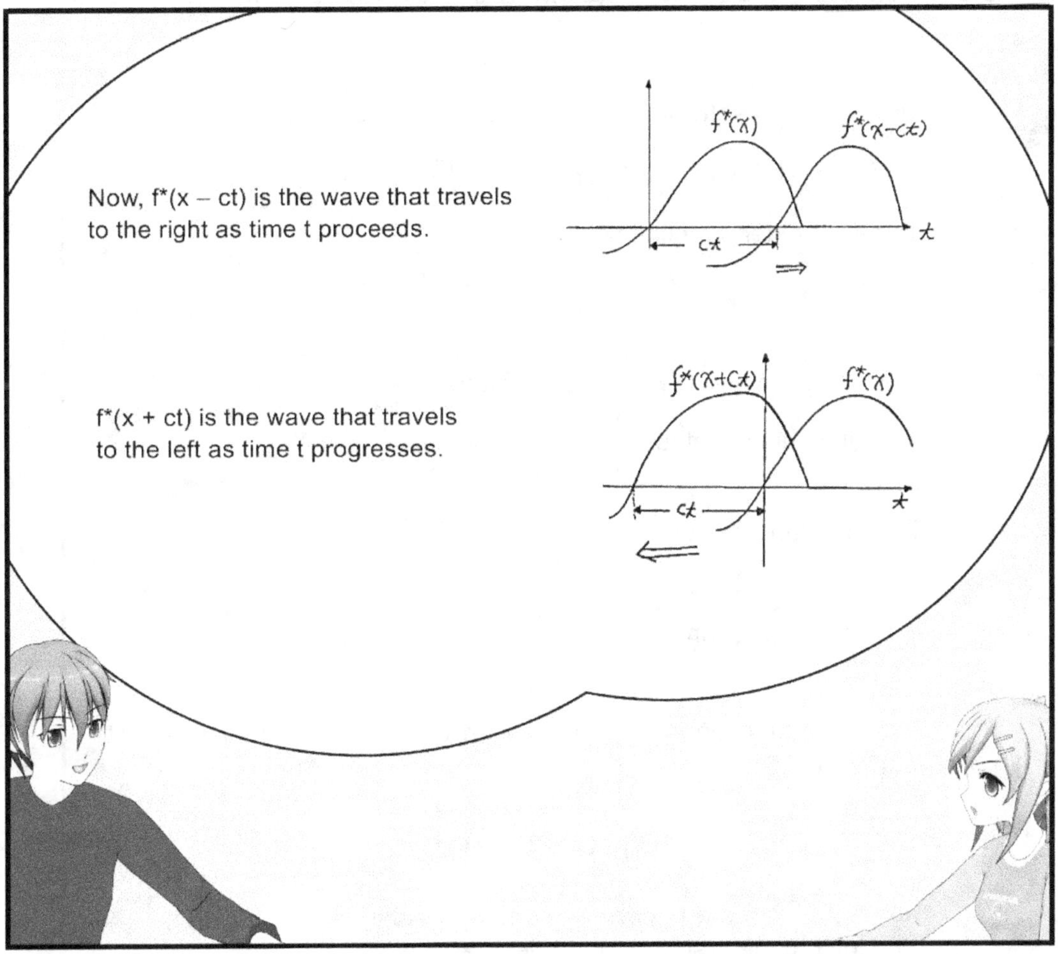

PATIAL DIFFERENTIAL EQUATIONS

Now, $f^*(x - ct)$ is the wave that travels to the right as time t proceeds.

$f^*(x + ct)$ is the wave that travels to the left as time t progresses.

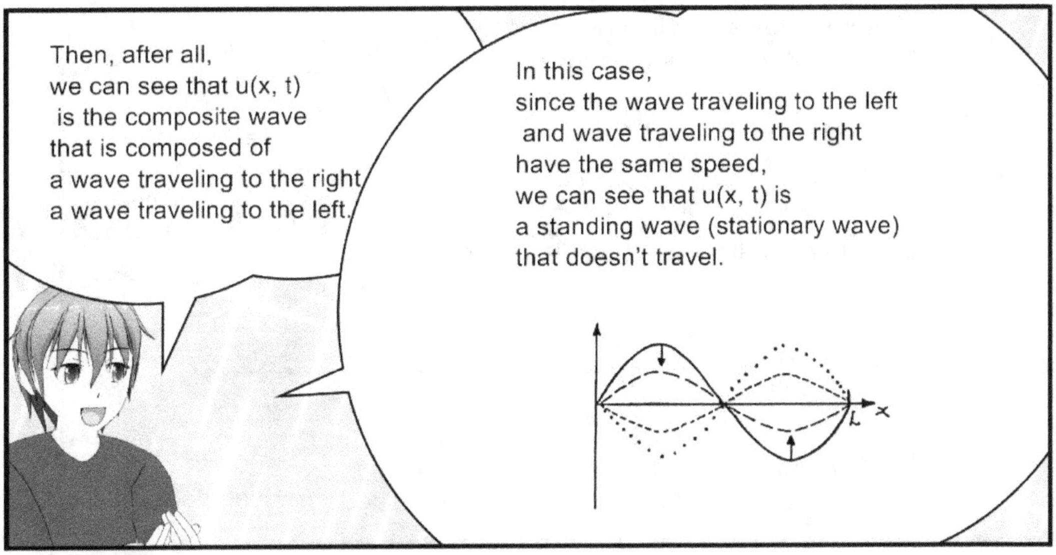

Then, after all, we can see that $u(x, t)$ is the composite wave that is composed of a wave traveling to the right, a wave traveling to the left.

In this case, since the wave traveling to the left and wave traveling to the right have the same speed, we can see that $u(x, t)$ is a standing wave (stationary wave) that doesn't travel.

Spices 1

Such solutions satisfying the boundary conditions as

$$u_n(x,t) = (B_n \cos\lambda_n t + B_n^* \sin\lambda_n t) \sin\frac{n\pi}{L}x \quad (n=1,2,3,\cdots)$$

are called eigen functions.

Where

$$\lambda_n = \frac{cn\pi}{L} = \sqrt{\frac{T}{\rho}}\frac{n\pi}{L} \quad (n=1,2,3,\cdots)$$

are called eigen values.

The set of eigen values

$$\{\lambda_1, \lambda_2, \cdots\}$$

is called a spectrum.

The eigen function is a harmonic vibration (a motion time-invariant) where the frequency is

$$\frac{\lambda_n}{2\pi} = \frac{cn}{2L} = \sqrt{\frac{T}{\rho}}\cdot\frac{n}{2L} \quad (n=1,2,3\cdots)$$

and is called the n-th normal mode of a string.
In addition, the first normal mode (n = 1) is called a fundamental mode, and the second and higher modes are called overtones.
In music, these are the octaves.

On the other hand, since $\sin\frac{n\pi x}{L}$ is 0 at

$$x = \frac{L}{n}, \frac{2L}{n}, \cdots, \frac{n-1}{n}L$$

the n-th normal mode has (n – 1) nodes.

> So, in case where both ends are fixed like guitar strings, only the vibrations in such normal modes as above are allowed.

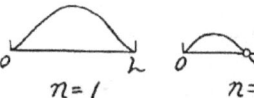

| Fundamental mode | The second normal mode | The third normal mode | The fourth normal mode |

PATIAL DIFFERENTIAL EQUATIONS

Quiz 1 How does the synthesizer make the sounds of a variety of musical instruments?

Answer

As stated earlier, the eigen functions

$$u_n(x,t) = (B_n \cos\lambda_n t + B_n^* \sin\lambda_n t) \sin\frac{n\pi}{L}x \quad (n=1,2,3,\cdots)$$

satisfying the boundary condition indicate the proper vibrations of a string fixed at x = 0 and x = L; the arbitrary wave pattern of a string is indicated by

$$\sum_{n=1}^{\infty} u_n(x,t)$$

the linear combination of eigen functions.
A synthesizer, electronic organ, etc. make a particular sound by means of the sum of appropriate choices of eigen functions.

In other words, the desired sound is composed of the basic sounds each of which corresponds to each of the terms in the Fourier series.

Quiz 2   What do we adjust when we tune the guitar strings?

Answer   Since the height of the sound is the height of the frequency, the tension of the string is adjusted.

Since the frequency of the fundamental mode is

$$\sqrt{\frac{T}{\rho}}\,\frac{1}{2L}$$

if we increase the tension T, the frequency gets higher
and the higher pitched sound gets produced.

In addition, since the frequency is in inverse proportion to the length L,
if we shorten the length, the higher pitched sound gets produced.

For instance, we know if we just strike the number 1 string,
the mi sound gets produced, but if we put the finger on the third fret
so that the length between both ends gets shorter,
the higher pitched sol sound gets produced, don't we?

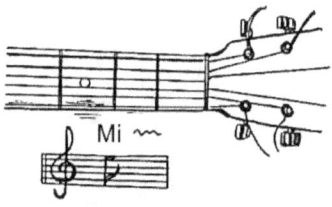
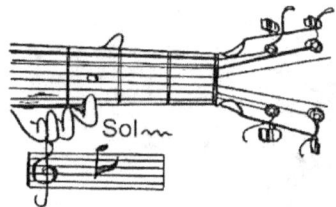

PATIAL DIFFERENTIAL EQUATIONS

Quiz 3

If we pull a string of length L where both ends are fixed so that it makes a triangle as shown below and let go of it, how is the string going to vibrate?

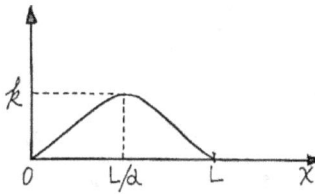

Answer

If we solve the 1-D wave equation and find the solution u(x, t) indicating the displacement in the string, we can see the way the string vibrates.

Set up the coordinate system as below.

Then, we need to find the solution u(x, t) satisfying the initial conditions that the initial displacement f(x) is

$$f(x) = \begin{cases} \frac{2k}{L} x & (0 < x < \frac{L}{2}) \\ \frac{2k}{L}(L-x) & (\frac{L}{2} < x < L) \end{cases}$$

and the initial velocity g(x) = 0.

The solution to the 1-D wave equation is

$$u(x, t) = \sum_{n=1}^{\infty}(B_n \cos\lambda_n t + B_n^* \sin\lambda_n t)\sin\frac{n\pi x}{L}$$

$$\left(\begin{array}{l} B_n = \frac{2}{L}\int_0^L f(x)\sin\frac{n\pi x}{L}dx \\ B_n^* = \frac{2}{cn\pi}\int_0^L g(x)\sin\frac{n\pi x}{L}dx \end{array} \right)$$

as found earlier.
In this case, since the initial velocity g(x) = 0, B*n = 0.

PARTIAL DIFFERENTIAL EQUATIONS

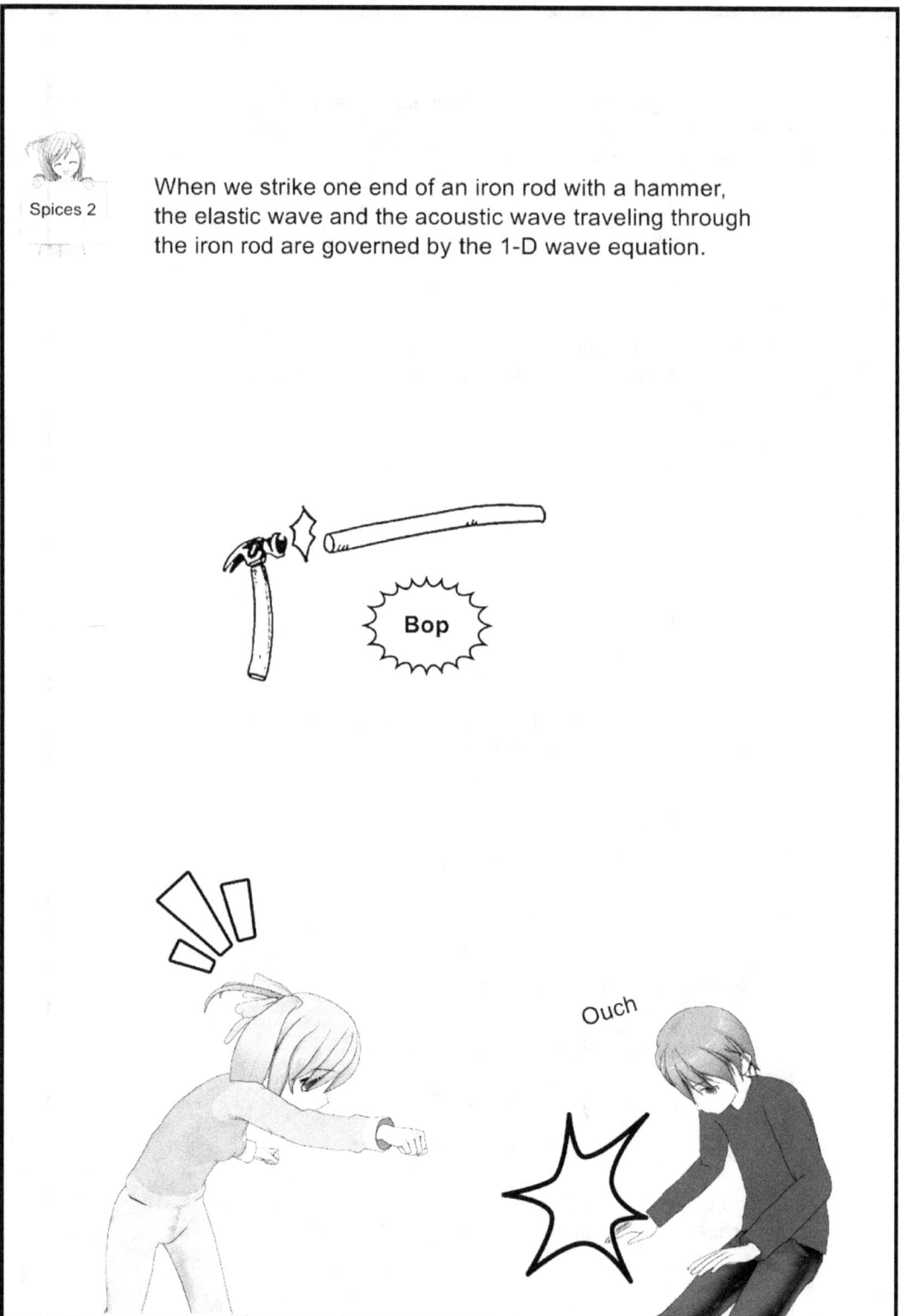

Spices 2: When we strike one end of an iron rod with a hammer, the elastic wave and the acoustic wave traveling through the iron rod are governed by the 1-D wave equation.

Bop

Ouch

PATIAL DIFFERENTIAL EQUATIONS

Spices 3

The small vertical vibration of the uniform beam
(for instance, a girder or the ironing board) as shown
in the figure is indicated by the 4th order PDE.

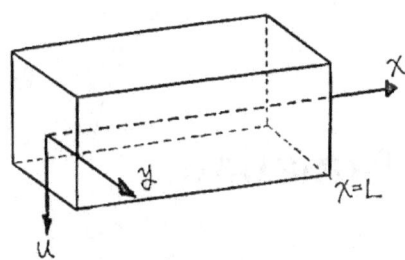

$$\frac{\partial^2 u}{\partial t^2} = -c^2 \frac{\partial^4 u}{\partial x^4}$$

$$\left(c^2 = \frac{EI}{\rho A} \right)$$

In this case, E is the Young's modulus, I is the moment of inertia
of the cross section with respect to y-axis, rho is the density,
and A is the cross sectional area.
Note that since the vertical vibration only is considered,
the eigen function (solution) is indicated by u(x, t), a function of
x and t only.

The fundamental difference between the beam (4th order PDE)
and the string (2nd order PDE) is that the beam is hardly bent
since the beam has rigidity. When we derive the wave equation
for the vibration of the string, we assume that the resistance against
bending doesn't exist. However, in case of the vibration of the beam,
we cannot allow such an assumption since it's unrealistic.
Consequently, the vibration of the beam is indicated by the 4th order
PDE unlike the vibration of the string.

If the beam is fixed as shown below, the boundary condition
is indicated by

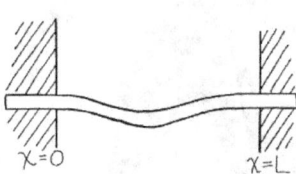

$$u(0, t) = 0, \quad u(L, t) = 0,$$
$$u_x(0, t) = 0, \quad u_x(L, t) = 0$$

The slopes at both ends
in the direction of the x-axis are 0,
which means that the ends are flat.

PARTIAL DIFFERENTIAL EQUATIONS

Modeling of vibrating membrane - 2-D wave equation

PARTIAL DIFFERENTIAL EQUATIONS

Since the 1-D wave equation indicating the vibration of the string was

$$\frac{\partial^2 u}{\partial t^2} = c^2 \frac{\partial^2 u}{\partial x^2}$$

the 2-D wave equation indicating the vibration of the membrane should be

$$\frac{\partial^2 u}{\partial t^2} = c^2 \left(\frac{\partial^2 u}{\partial x^2} + \frac{\partial^2 u}{\partial y^2} \right)$$

Diffeq! Your presently listening to a sound, that is, the vibration of your eardrum can also be put in the 2-D wave equation.

All right! Now, how do we solve the 2-D wave equation?

The solution to the 2-D wave equation has 3 variables x, y, and t, and should be u(x, y, t).

So, we separate the PDE into 3 ODEs by the separation of variables, and then, solve the ODEs!

Let's actually solve the equation for the drumhead of a rectangular drum, then.

PATIAL DIFFERENTIAL EQUATIONS

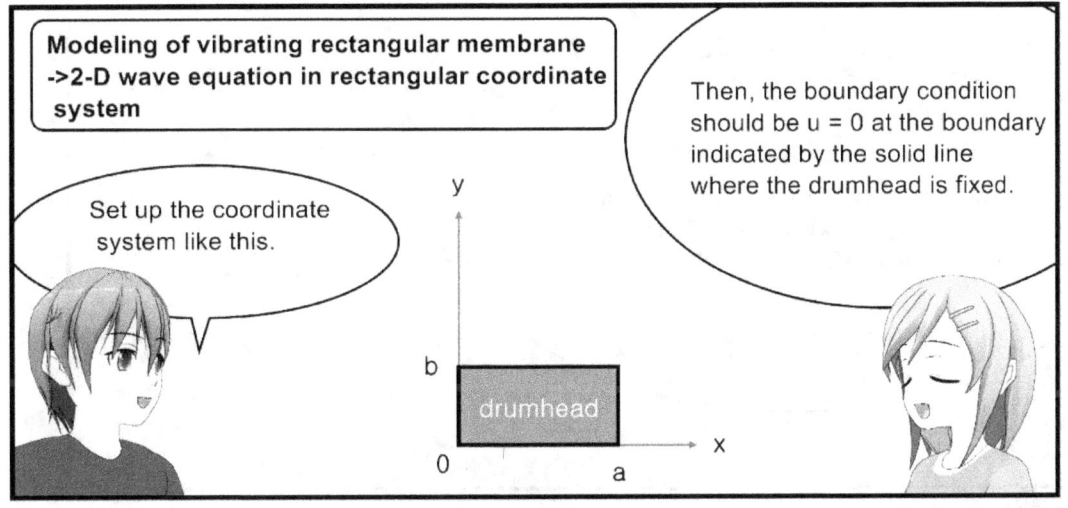

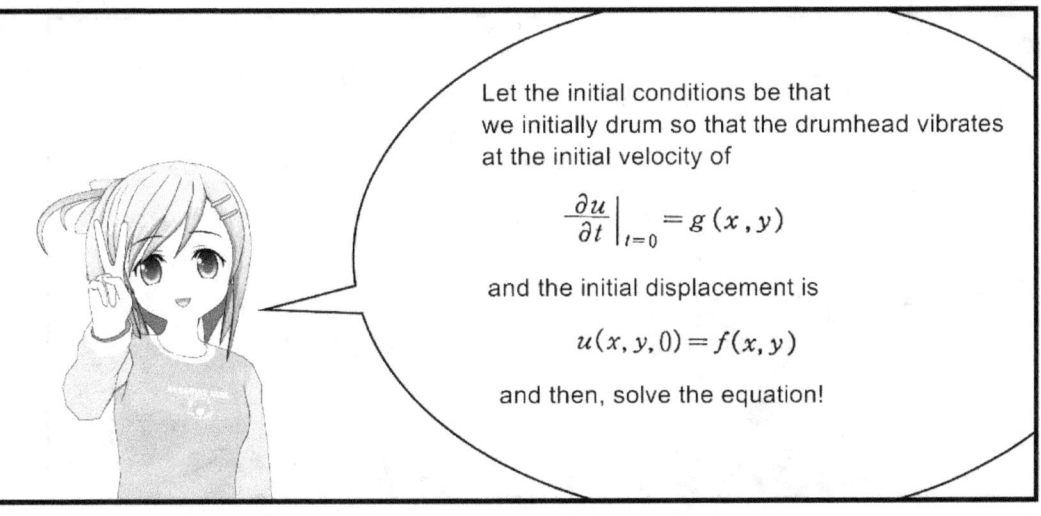

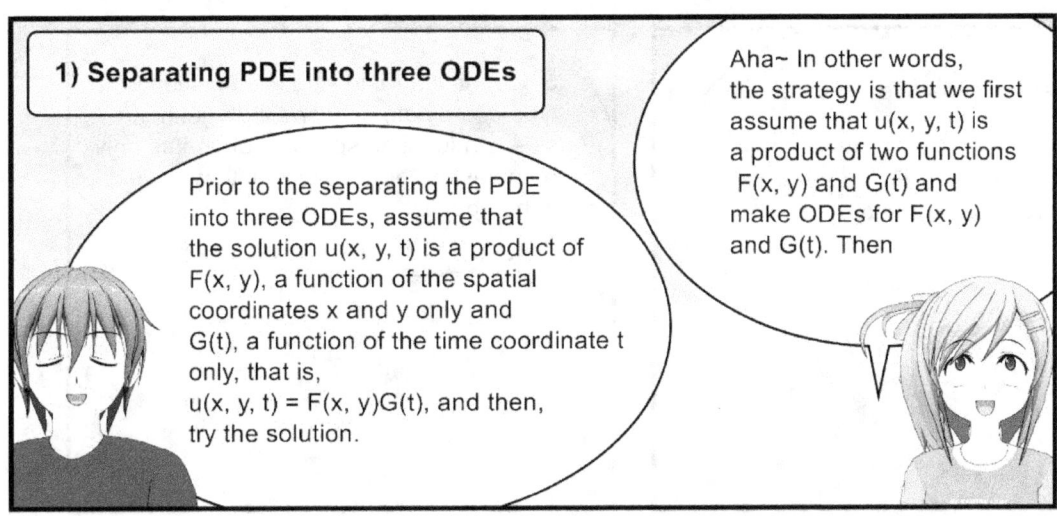

PARTIAL DIFFERENTIAL EQUATIONS

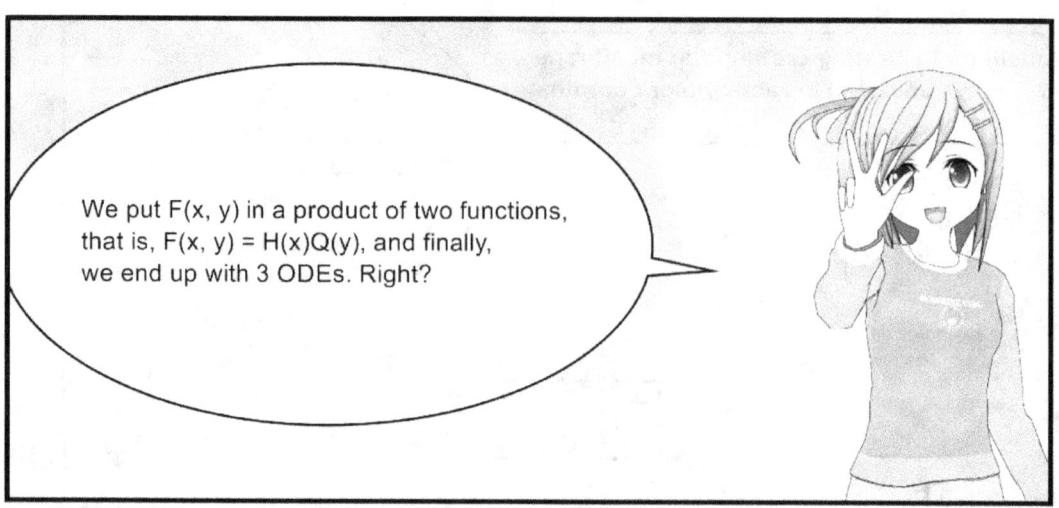

We put F(x, y) in a product of two functions, that is, F(x, y) = H(x)Q(y), and finally, we end up with 3 ODEs. Right?

Put u(x, y, t) = F(x, y)G(t) into the 2-D wave equation

$$\frac{\partial^2 u}{\partial t^2} = c^2 \left(\frac{\partial^2 u}{\partial x^2} + \frac{\partial^2 u}{\partial y^2} \right)$$

After the substitution, we get*

$$F\ddot{G} = c^2(F_{xx}G + F_{yy}G)$$

$$F_{xx} = \frac{\partial^2 F}{\partial x^2}$$

After the separation of variables, we get

$$\frac{\ddot{G}}{c^2 G} = \frac{F_{xx} + F_{yy}}{F}$$

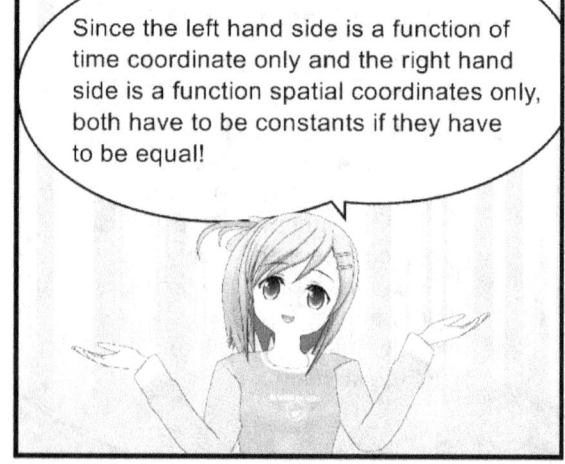

Since the left hand side is a function of time coordinate only and the right hand side is a function spatial coordinates only, both have to be constants if they have to be equal!

PATIAL DIFFERENTIAL EQUATIONS

Suppose the constant is k and consider the three cases where k = 0, k > 0, k < 0. Then, as in the case of the 1-D wave equation, we get a significant solution only if k < 0.

So, setting

$$k = -\nu^2$$

so that k is always negative, we get

$$\frac{\ddot{G}}{c^2 G} = \frac{F_{xx} + F_{yy}}{F} = -\nu^2$$

Then, we get two ODEs from the above.

$$\ddot{G} + c^2\nu^2 G = 0 \quad \boxed{\ddot{G} + \lambda^2 G = 0}$$

$$\boxed{F_{xx} + F_{yy} + \nu^2 F = 0}$$

where $\lambda = c\nu$, G is a time function, and F is an amplitude function (Helmholtz equation)

Next, we need to get the two ODEs from the Helmholtz equation

$$F_{xx} + F_{yy} + \nu^2 F = 0$$

Then, assume that F(x, y) is a product of two functions, that is, F(x, y) = H(x)Q(y), and put it into the PDE.

PARTIAL DIFFERENTIAL EQUATIONS

Then, we get*
$$H''Q + HQ'' + \nu^2 HQ = 0$$

By the separation of variables, we get
$$\frac{1}{H}H'' = -\frac{1}{Q}(Q'' + \nu^2 Q)$$

Since the left hand side is a function of x only and the right hand side is a function of y only, both need to be constants if they are equal.

As in the case of the 1-D wave equation, since we get a significant solution only if the constant is negative, setting the constant to be equal to $-k^2$

we get
$$\frac{1}{H}H'' = -\frac{1}{Q}(Q'' + \nu^2 Q) = -k^2$$

Then, from the above, we get the two ODEs as below:
$$H'' + k^2 H = 0$$
$$Q'' + p^2 Q = 0 \quad (p^2 = \nu^2 - k^2)$$

Now, we have the three ODEs
$$\begin{cases} \ddot{G} + \lambda^2 G = 0 & (\lambda = c\nu) \\ H'' + k^2 H = 0 \\ Q'' + p^2 Q = 0 & (p^2 = \nu^2 - k^2) \end{cases}$$

음... 모두 같은 형태네.

PATIAL DIFFERENTIAL EQUATIONS

Finding the general solution to the above, we get

$$\begin{cases} H(x) = A\cos kx + B\sin kx \\ Q(y) = C\cos py + D\sin py \\ G(t) = B\cos \lambda t + B^*\sin \lambda t \end{cases}$$

2) Satisfying boundary conditions

The second thing we have to do is that we get the general solution to satisfy the boundary conditions, isn't it?

Since the boundary conditions have to do with spatial coordinates only, we get the amplitude function $F(x, y) = H(x)Q(y)$ to be 0 at the lines where the drumhead is fixed indicated by $x = 0$, $x = a$, $y = 0$, and $y = b$!

Putting the boundary conditions in expressions, we get

$$F(0,y) = 0, \quad F(a,y) = 0$$
$$F(x,0) = 0, \quad F(x,b) = 0$$

Consider them in tern.

i) $F(0, y) = 0$

$$F(0, y) = H(0) Q(y) = 0$$

So,
$H(0) = 0$ or $Q(y) = 0$.

PATIAL DIFFERENTIAL EQUATIONS

Therefore, it has to be that

$ka = m\pi \quad (m = 1, 2, \cdots)$

$k = \dfrac{m\pi}{a} \quad (m = 1, 2, \cdots)$

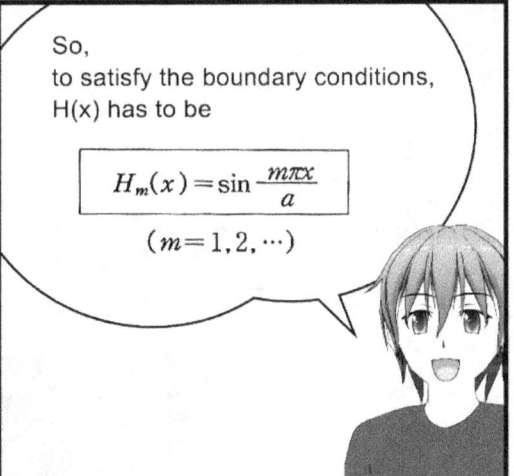

So, to satisfy the boundary conditions, $H(x)$ has to be

$$\boxed{H_m(x) = \sin \dfrac{m\pi x}{a}}$$

$(m = 1, 2, \cdots)$

For the same reason, from

iii) $F(x, 0) = 0$,
iv) $F(x, b) = 0$

$F(x, 0) = H(x) Q(0) = 0$
$F(x, b) = H(x) Q(b) = 0$

Consequently,

$Q(0) = C = 0$
$Q(b) = D \sin pb = 0$

$\sin pb = 0$

$p = \dfrac{n\pi}{b} \quad (n = 1, 2, \cdots)$

Therefore,

$$\boxed{Q_n(y) = \sin \dfrac{n\pi y}{b}}$$

$(n = 1, 2, \cdots)$

From the above, $F = HQ$ is

$$\boxed{\begin{aligned} F_{mn}(x, y) &= H_m(x) Q_n(y) \\ &= \sin \dfrac{m\pi x}{a} \sin \dfrac{n\pi y}{b} \end{aligned}}$$

$m = 1, 2, \cdots$
$n = 1, 2, \cdots$

which is the amplitude function satisfying the boundary conditions!

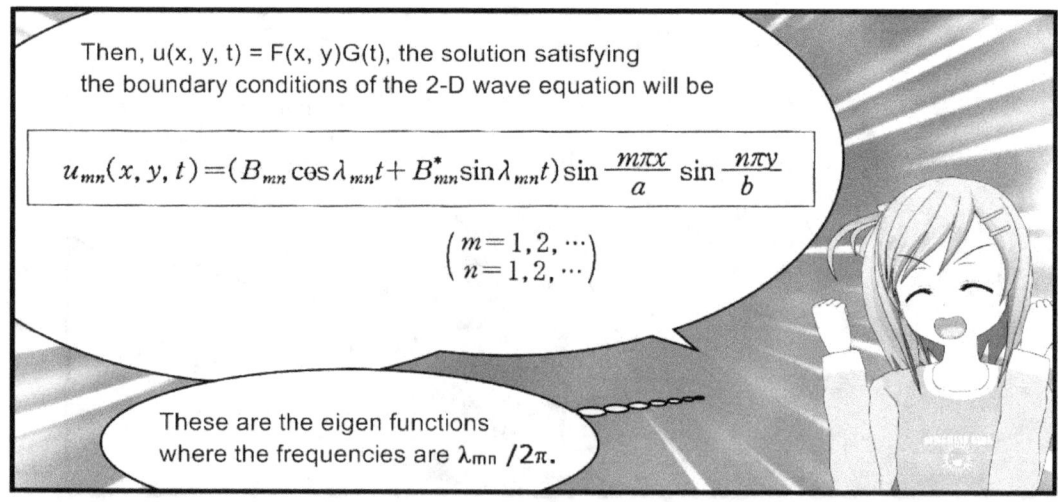

PATIAL DIFFERENTIAL EQUATIONS

Finally, we need to satisfy the initial conditions.

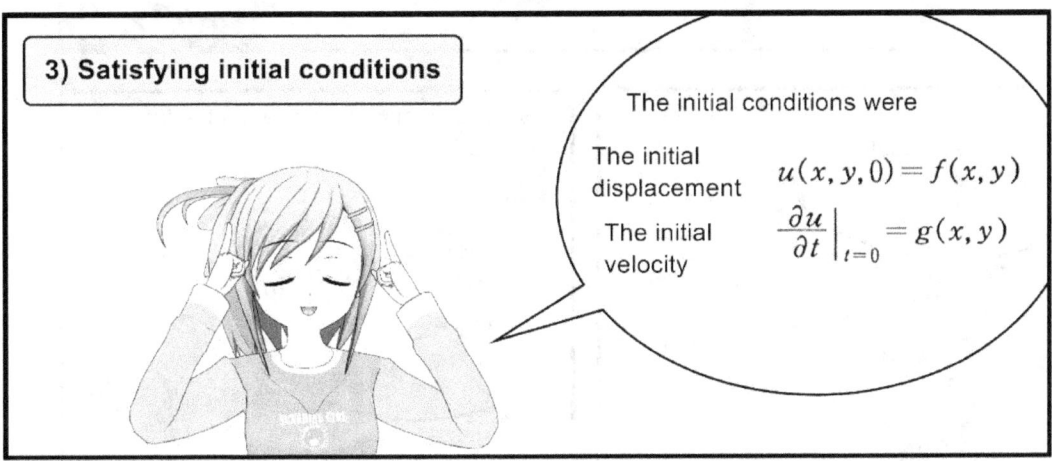

3) Satisfying initial conditions

The initial conditions were

The initial displacement $u(x, y, 0) = f(x, y)$

The initial velocity $\left.\dfrac{\partial u}{\partial t}\right|_{t=0} = g(x, y)$

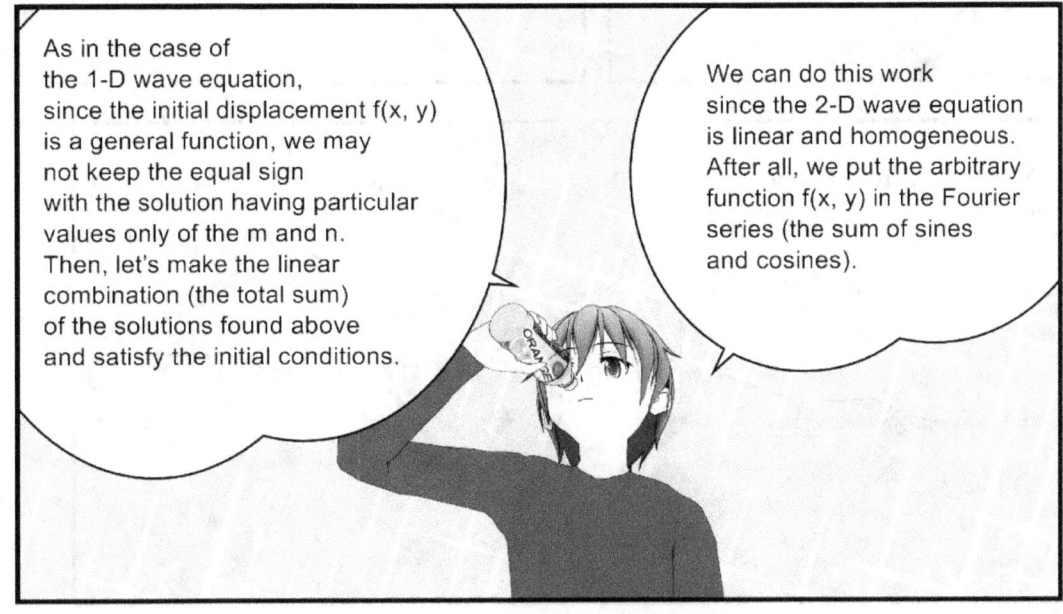

As in the case of the 1-D wave equation, since the initial displacement f(x, y) is a general function, we may not keep the equal sign with the solution having particular values only of the m and n. Then, let's make the linear combination (the total sum) of the solutions found above and satisfy the initial conditions.

We can do this work since the 2-D wave equation is linear and homogeneous. After all, we put the arbitrary function f(x, y) in the Fourier series (the sum of sines and cosines).

However, in case of the 2-D wave equation, we need to take the total sum for the two indices m and n, and there will be two of sigmas. Right? That is,

$$u(x, y, t) = \sum_{m=1}^{\infty} \sum_{n=1}^{\infty} u_{mn}(x, y, t)$$
$$= \sum_{m=1}^{\infty} \sum_{n=1}^{\infty} (B_{mn} \cos \lambda_{mn} t + B_{mn}^* \sin \lambda_{mn} t) \cdot \sin \frac{m\pi x}{a} \sin \frac{n\pi y}{b}$$

Now, will you apply the initial conditions?

Since the initial displacement is f(x, y), we get

$$u(x, y, 0)$$
$$= \sum_{m=1}^{\infty} \sum_{n=1}^{\infty} B_{mn} \sin \frac{m\pi x}{a} \sin \frac{n\pi y}{b}$$
$$= f(x, y)$$

That's called a double Fourier series! Setting

$$K_m(y) = \sum_{n=1}^{\infty} B_{mn} \sin \frac{n\pi y}{b}$$

we get

$$f(x, y) = \sum_{m=1}^{\infty} K_m(y) \sin \frac{m\pi x}{a}$$

As in the case of the initial displacement,
if we apply the double Fourier series, we get

$$B^*_{mn} = \frac{4}{ab\lambda_{mn}} \int_0^b \int_0^a g(x,y) \sin\frac{m\pi x}{a} \sin\frac{n\pi y}{b}\, dx\, dy$$

$$\begin{pmatrix} m = 1, 2, \cdots \\ n = 1, 2, \cdots \end{pmatrix}$$

Summing up, the solution satisfying the rectangular boundary condition along with the initial condition of the 2-D wave equation

$$\frac{\partial^2 u}{\partial t^2} = c^2 \left(\frac{\partial^2 u}{\partial x^2} + \frac{\partial^2 u}{\partial y^2} \right)$$

is

$$u(x,y,t) = \sum_{m=1}^{\infty} \sum_{n=1}^{\infty} (B_{mn}\cos\lambda_{mn}t + B^*_{mn}\sin\lambda_{mn}t)\sin\frac{m\pi x}{a}\sin\frac{n\pi y}{b}$$

$$\begin{pmatrix} B_{mn} = \dfrac{4}{ab} \int_0^b \int_0^a f(x,y)\sin\dfrac{m\pi x}{a}\sin\dfrac{n\pi y}{b}\, dx\, dy \\ B^*_{mn} = \dfrac{4}{ab\lambda_{mn}} \int_0^b \int_0^a g(x,y)\sin\dfrac{m\pi x}{a}\sin\dfrac{n\pi y}{b}\, dx\, dy \end{pmatrix}$$

$$\lambda_{mn} = c\pi\sqrt{\frac{m^2}{a^2} + \frac{n^2}{b^2}} \quad \begin{pmatrix} m = 1, 2, \cdots \\ n = 1, 2, \cdots \end{pmatrix}$$

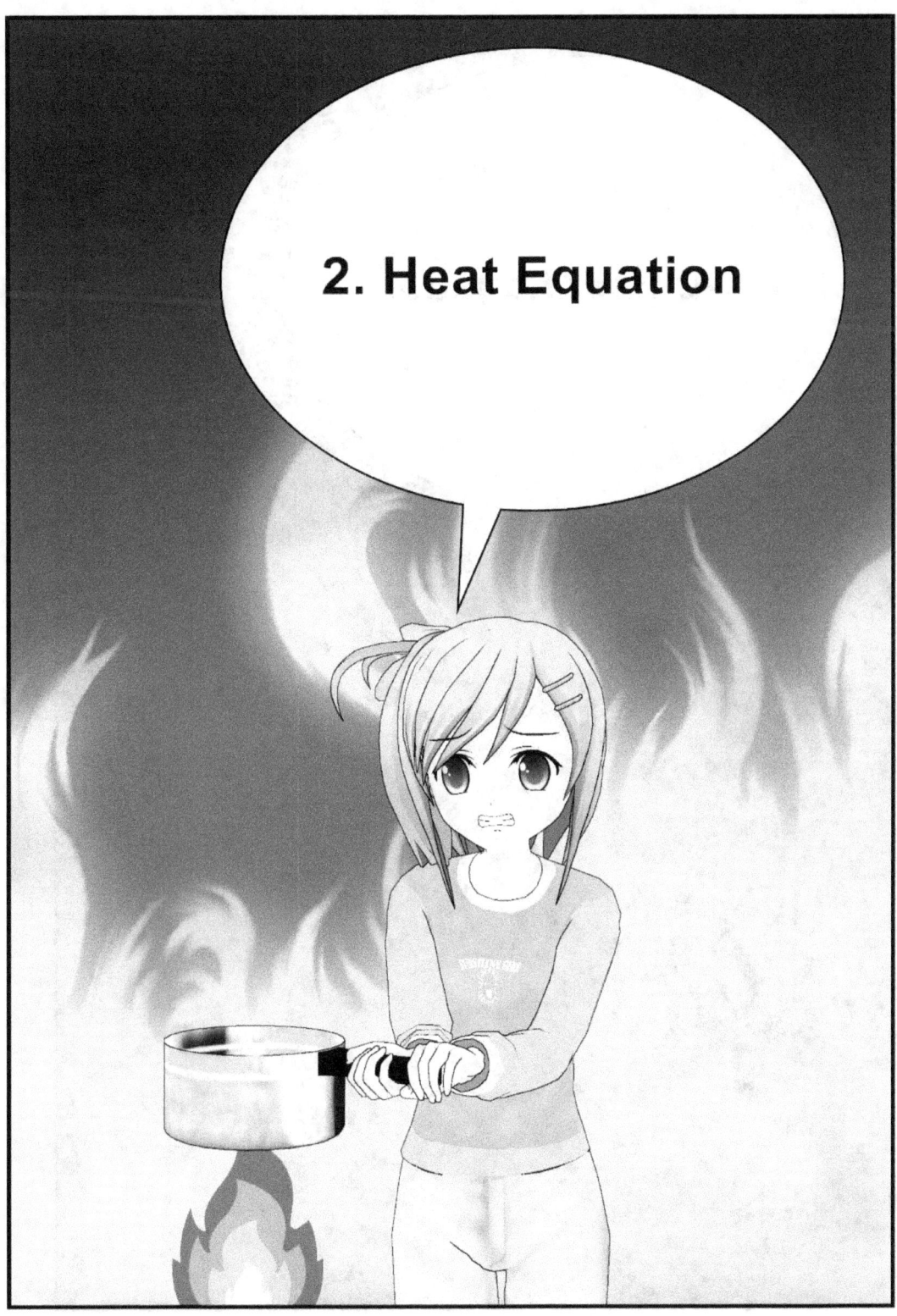

PATIAL DIFFERENTIAL EQUATIONS

Modeling of heat conduction inside a long and thin rod - 1-D heat equation

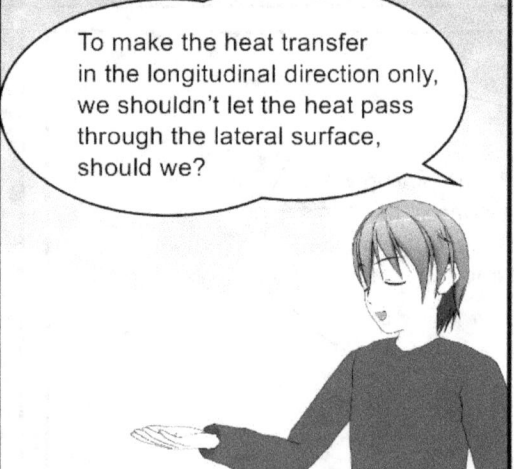

PATIAL DIFFERENTIAL EQUATIONS

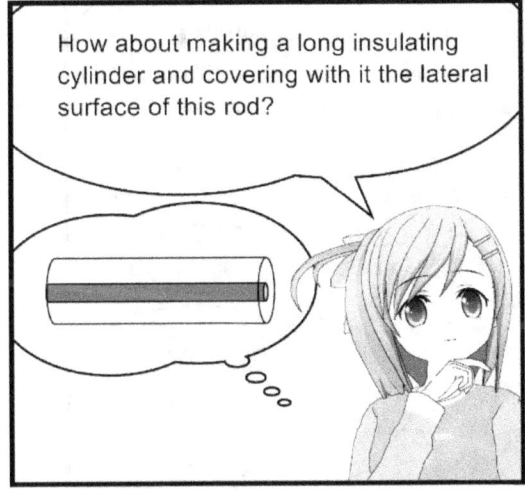

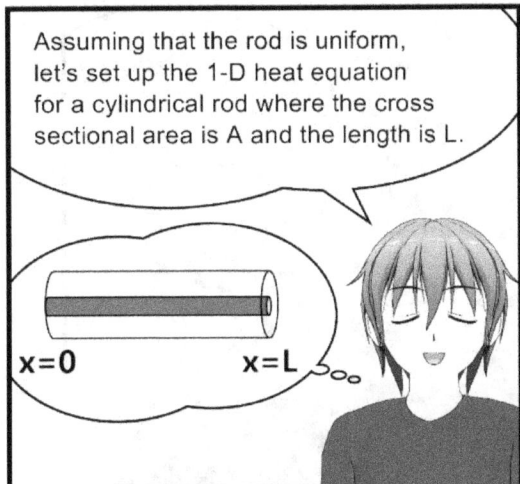

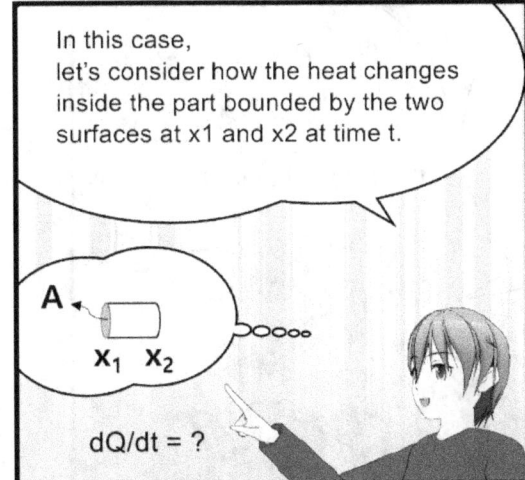

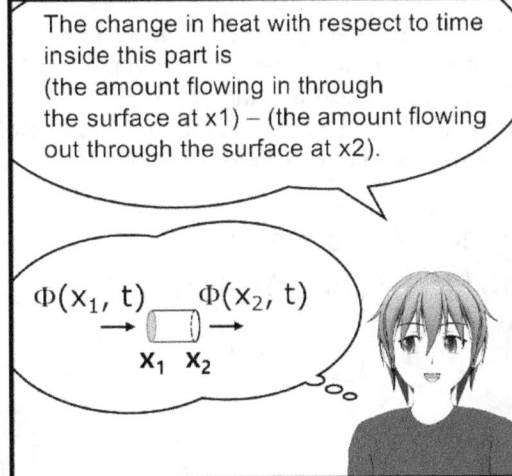

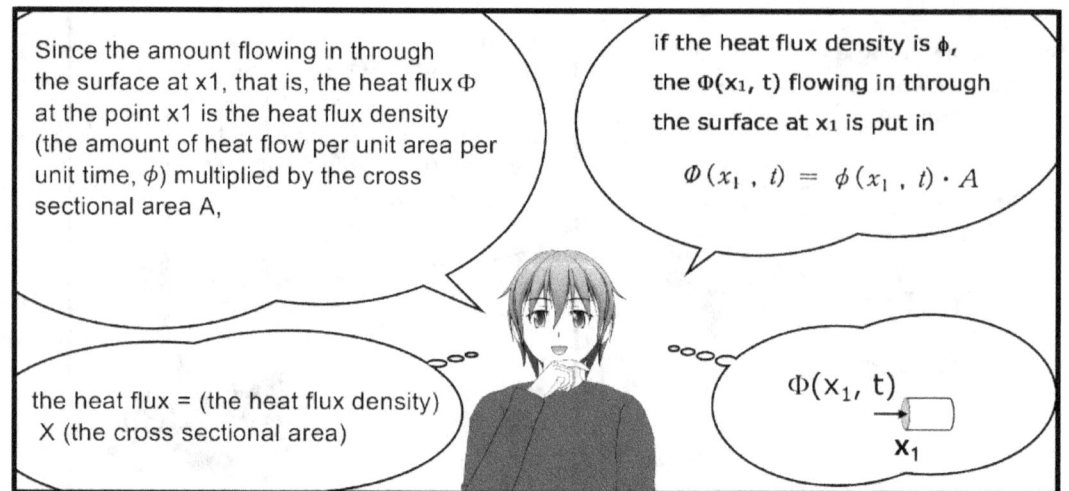

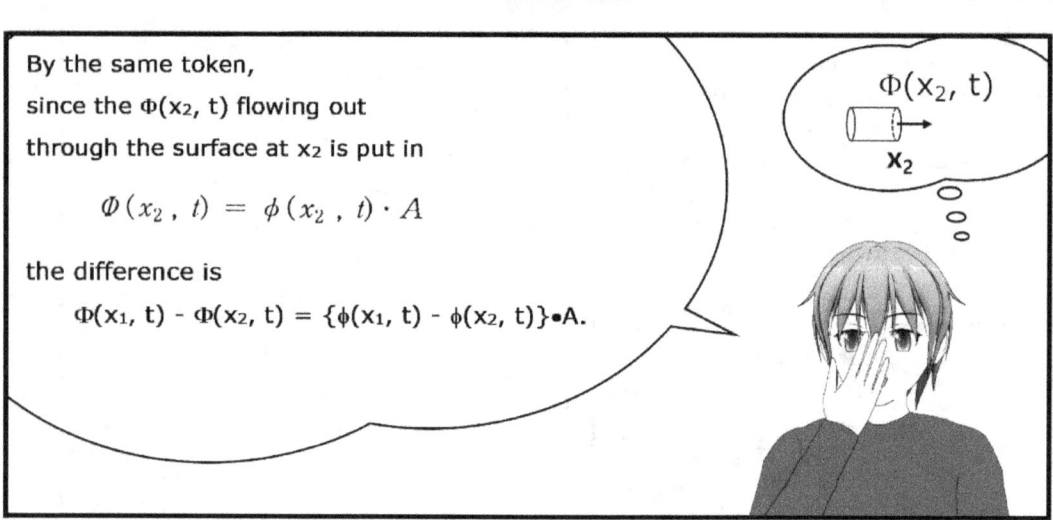

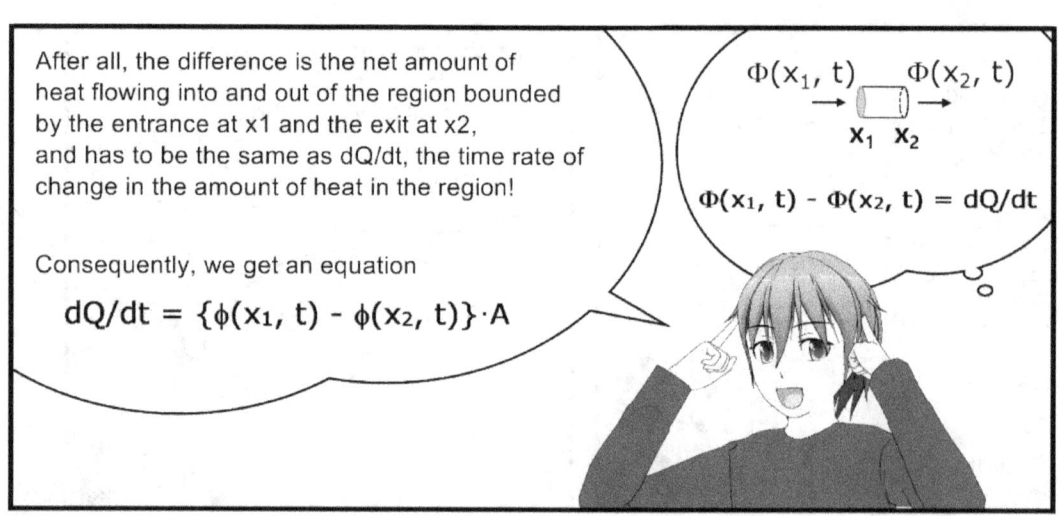

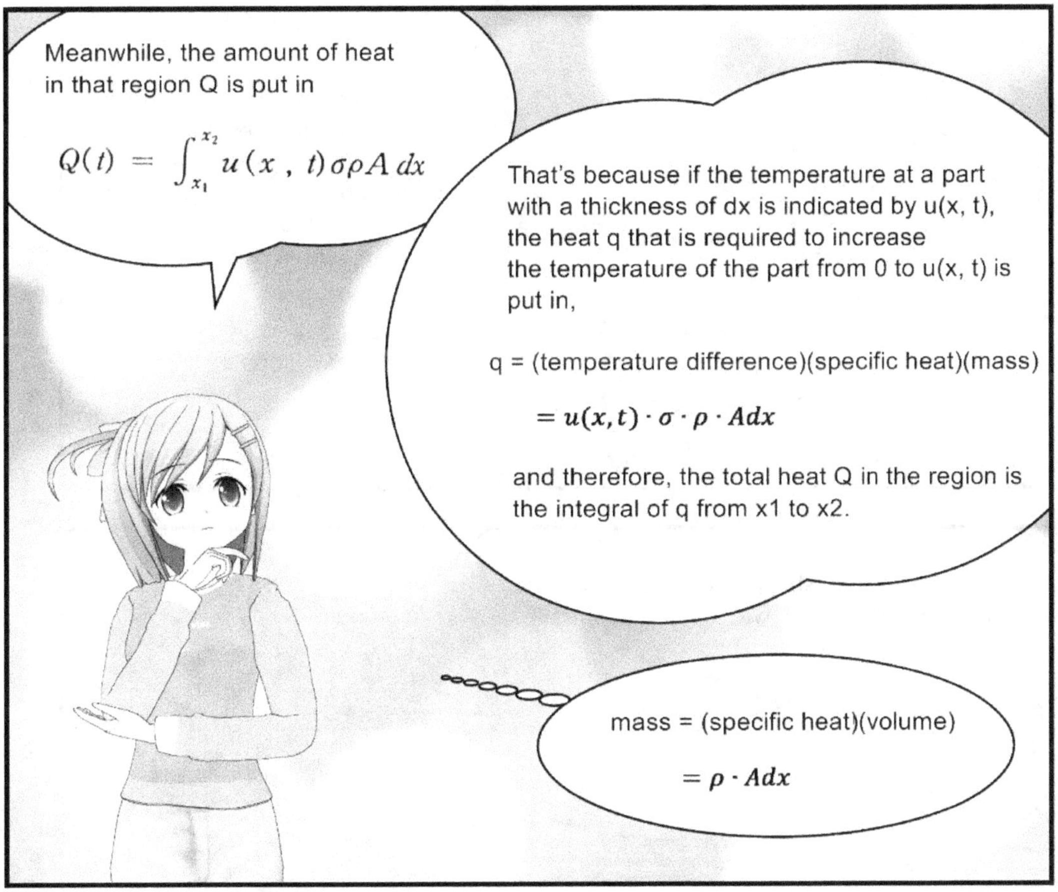

PATIAL DIFFERENTIAL EQUATIONS

Meanwhile, the amount of heat in that region Q is put in

$$Q(t) = \int_{x_1}^{x_2} u(x, t)\sigma\rho A\, dx$$

That's because if the temperature at a part with a thickness of dx is indicated by u(x, t), the heat q that is required to increase the temperature of the part from 0 to u(x, t) is put in,

q = (temperature difference)(specific heat)(mass)

$$= u(x,t) \cdot \sigma \cdot \rho \cdot A dx$$

and therefore, the total heat Q in the region is the integral of q from x1 to x2.

mass = (specific heat)(volume)

$$= \rho \cdot A dx$$

So, the time rate of change of the Q will be

$$\frac{dQ}{dt} = \int_{x_1}^{x_2} \sigma\rho \frac{\partial u(x,t)}{\partial t} A\, dx$$

(σ, ρ, A: constant)

Putting it into the equation above, we get

$$\{\phi(x_1, t) - \phi(x_2, t)\}A$$

$$= \int_{x_1}^{x_2} \sigma\rho \frac{\partial u(x,t)}{\partial t} \cdot A \cdot dx$$

the heat flux density ϕ is proportional to the temperature gradient

$$\phi(x, t) = -K\frac{\partial u(x, t)}{\partial x}$$

In this case, k is the heat conductivity!

Since heat flows from that point at the higher temperature to the point at the lower temperature but the direction of the gradient is from the lower to the higher, (−) sign is put.

Putting the relationship into the last equation in the preceding page, we get

$$K\frac{\partial u(x_2, t)}{\partial x} - K\frac{\partial u(x_1, t)}{\partial x} = \int_{x_1}^{x_2} \sigma\rho \frac{\partial u(x, t)}{\partial t} dx$$

In the above, since the left hand side is equal to

$$\int_{x_1}^{x_2} \frac{\partial}{\partial x}\left[K\frac{\partial u(x, t)}{\partial x} \right] dx$$

putting this into the left hand side, we get

$$\int_{x_1}^{x_2} \frac{\partial}{\partial x}\left[K\frac{\partial u(x, t)}{\partial x} \right] dx = \int_{x_1}^{x_2} \sigma\rho \frac{\partial u(x, t)}{\partial t} dx$$

PATIAL DIFFERENTIAL EQUATIONS

PARTIAL DIFFERENTIAL EQUATIONS

Spices 1   The diffusion of material inside fluid (gas or liquid) or solid is governed by the same equation as the heat equation, too.

For instance:

1 If we fill up a thin glass tube with water
and drop a drop of red ink into the tube,
the ink diffuses into the water.
Now, if we assume that the ink diffuses
in the longitudinal direction only in the tube,
the diffusion is one-dimensional and
is governed by

$$\frac{\partial u}{\partial t} = c^2 \frac{\partial^2 u}{\partial x^2}$$

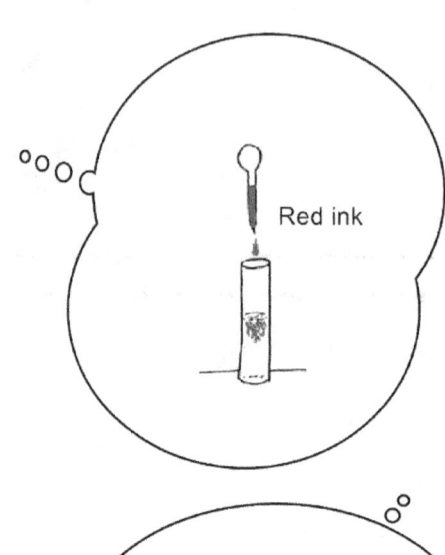

in the same form of the 1-D heat equation.
In this case, u is the concentration of the ink.

2. If we put together an aluminum rod
and a copper rod longitudinally and heat them up
inside an electric furnace, the aluminum atoms
diffuse into the copper rod (the copper atoms
diffuse into the aluminum rod at the same time)
and the alloy of aluminum and copper is produced.
This process is also the 1-D diffusion
and governed by

$$\frac{\partial u}{\partial t} = c^2 \frac{\partial^2 u}{\partial x^2}$$

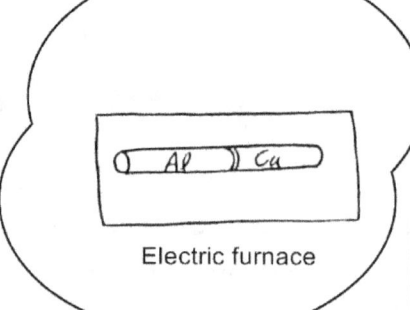

In this case, u is the concentration of aluminum.

3. When pollutant outlet from a chimney diffuses
into atmosphere, the diffusion is three-dimensional,
and is governed by

$$\frac{\partial u}{\partial t} = c^2 \nabla^2 u$$
$$\left(\nabla^2 u = \frac{\partial^2 u}{\partial x^2} + \frac{\partial^2 u}{\partial y^2} + \frac{\partial^2 u}{\partial z^2} \right)$$

In this case, u is the concentration of the pollutant.

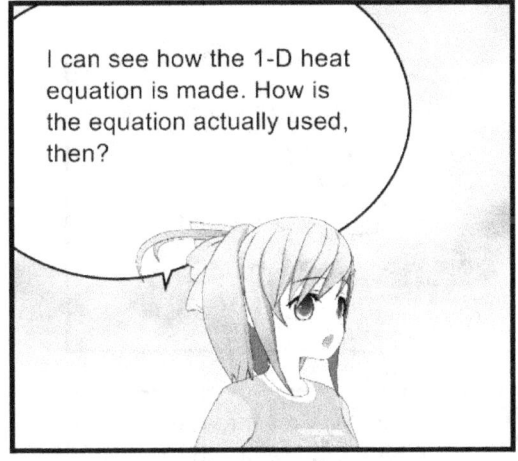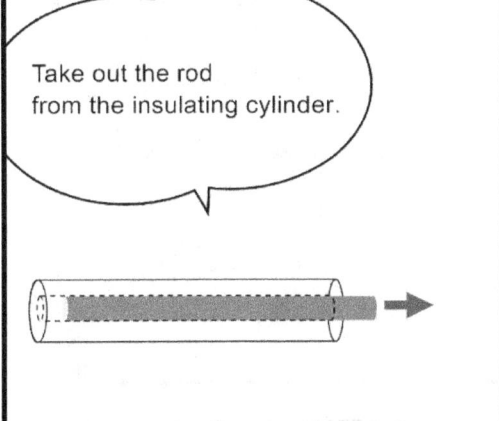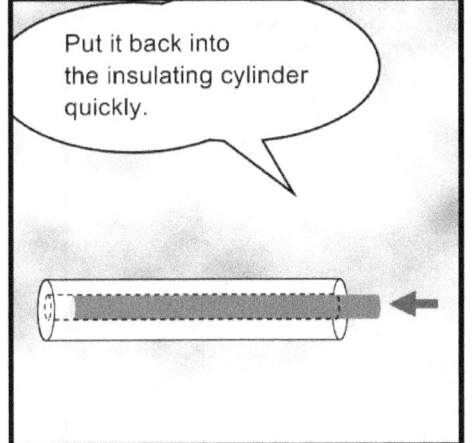

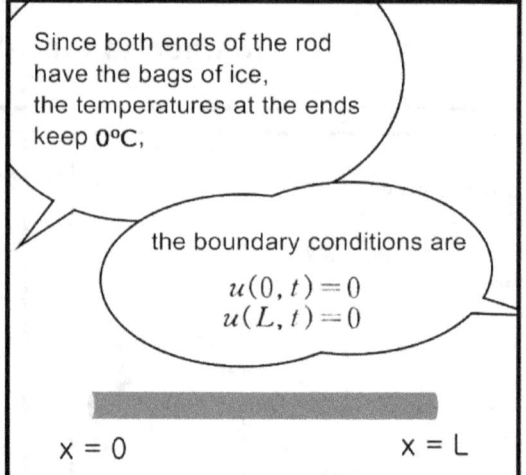
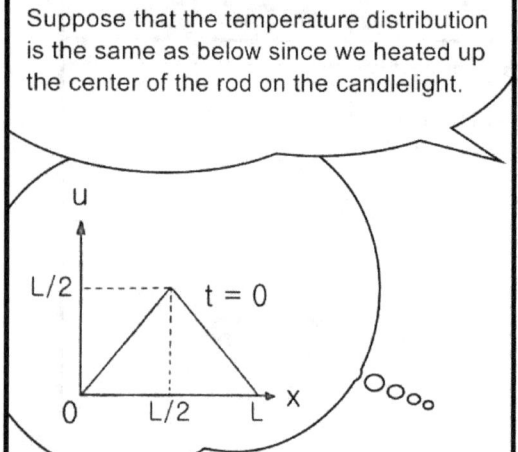

PATIAL DIFFERENTIAL EQUATIONS

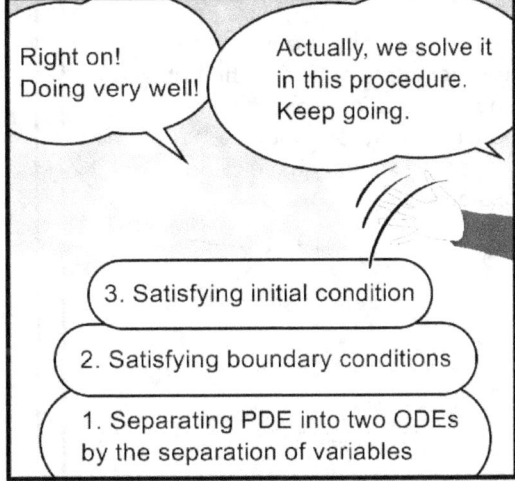

PARTIAL DIFFERENTIAL EQUATIONS

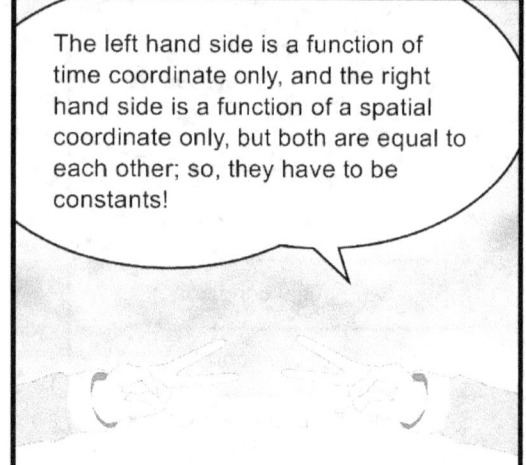

The left hand side is a function of time coordinate only, and the right hand side is a function of a spatial coordinate only, but both are equal to each other; so, they have to be constants!

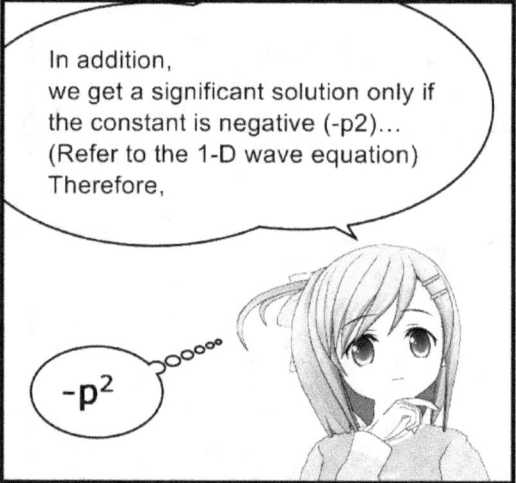

In addition, we get a significant solution only if the constant is negative ($-p^2$)... (Refer to the 1-D wave equation) Therefore,

$-p^2$

$$\frac{\dot{G}}{C^2 G} = \frac{F''}{F} = -p^2$$

From the above, we get the two ODEs as follows.

- $F'' + p^2 F = 0$
- $\dot{G} + C^2 p^2 G = 0$

Since the general solution to the first equation is

$F(x) = A \cos px + B \sin px$

the next thing to be done is to get this to satisfy the boundary conditions!

Don't know about how to find the general solution? Then, refer to the ODE Book!

PATIAL DIFFERENTIAL EQUATIONS

2. Satisfying boundary conditions

Satisfying the boundary conditions
$u(0, t) = 0$
$u(L, t) = 0$
we need to set

Remember the bags of ice?

$F(0)\,G(t) = 0$
$F(L)\,G(t) = 0$

Why? That's because

$u(x, t) = F(x)G(t)$

If $G(t)$ is 0, $u = 0$ and an insignificant solution results. So, it has to be that

$F(0) = 0$, $F(L) = 0$

Since
$F(x) = A \cos px + B \sin px$

$F(0) = A = 0$
$F(L) = B \sin pL = 0$

(if $B = 0$, $F = 0$ -> $u = 0$ -> nonsense)
Therefore,
$$\sin pL = 0$$
Consequently,
$$p = \frac{n\pi}{L} \quad (n = 1, 2, \cdots)$$

$A = 0$, $p = \frac{n\pi}{L}$, $B = 1$

If we put this into $F(x)$,

(Set $B = 1$ since a constant is multiplied in $G(t)$ anyway later)

$$F_n(x) = \sin \frac{n\pi x}{L} \quad (n = 1, 2, \cdots)$$

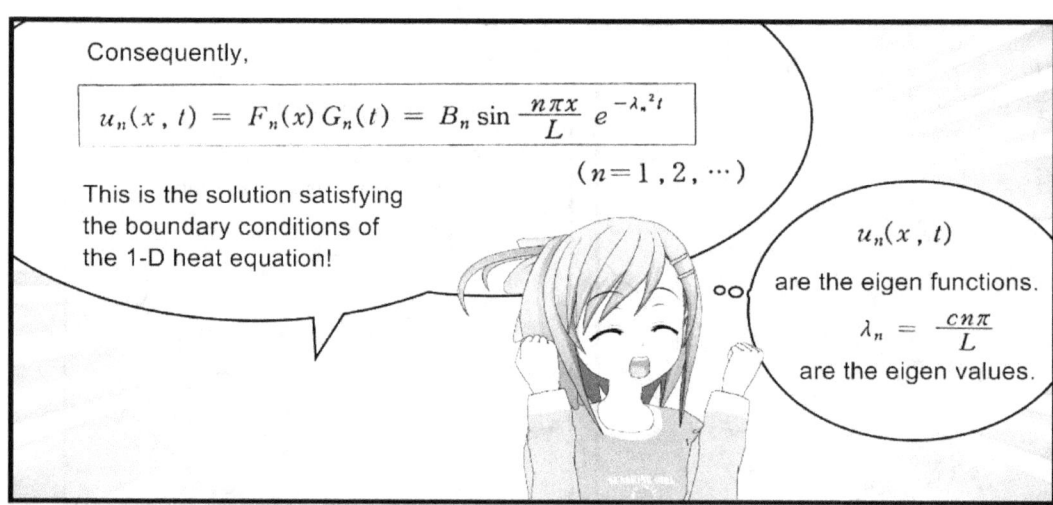

PATIAL DIFFERENTIAL EQUATIONS

Quiz 1

A long thin copper rod with length of 50 cm is insulated laterally and both ends are maintained at temperature zero degree Celsius. Find the temperature u(x, t) when initial temperature is

$$90\sin(\pi x/50) \,°C$$

Note thermal conductivity, specific heat, and density of copper are

$$k = 0.95 \; cal/cm \cdot sec \cdot °C, \; \sigma = 0.092 \; cal/g \cdot °C, \; \rho = 8.92 \; g/cm^3$$

Answer

Applying 1D heat equation, the solution of u(x,t) is given by

$$u(x,t) = \sum_{n=1}^{\infty} B_n \sin\frac{n\pi x}{L} e^{-\lambda n^2 t}$$

$$B_n = \frac{2}{L}\int_0^L f(x) \sin\frac{n\pi x}{2} dx \quad (n = 1,2,3,\cdots)$$

$$\lambda_n = \frac{cn\pi}{L}, \quad c = \sqrt{\frac{k}{\sigma\rho}}$$

Applying the initial condition

$$u(x,0) = \sum_{n=1}^{\infty} B_n \sin\frac{n\pi x}{50} e^{-\lambda n^2 \cdot 0} = 90\sin\frac{\pi x}{50}$$

Therefore

$$B_1 = 90, B_2 = B_3 = B_4 = \cdots = 0$$

And

$$\lambda_1^2 = \frac{c^2\pi^2}{L^2} = \frac{k}{\sigma\rho}\frac{\pi^2}{L^2}$$

$$= \frac{(0.95)}{(0.092)(8.92)}\frac{(3.14)^2}{(50)^2} \cong 0.0046 \; s^{-1}$$

Substituting these values into u(x,t), we get

$$u(x,t) = 90\sin\frac{\pi x}{50} e^{-0.0046 t}$$

PATIAL DIFFERENTIAL EQUATIONS

Modeling of heat conduction inside membrane -> 2-D heat equation

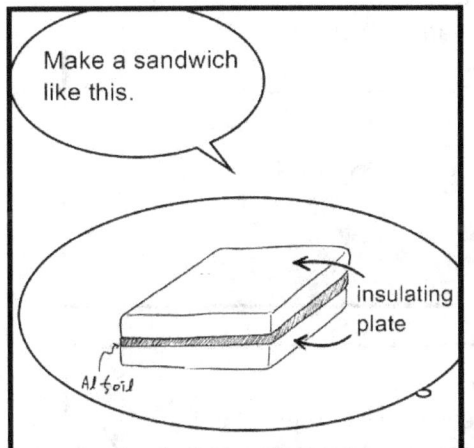

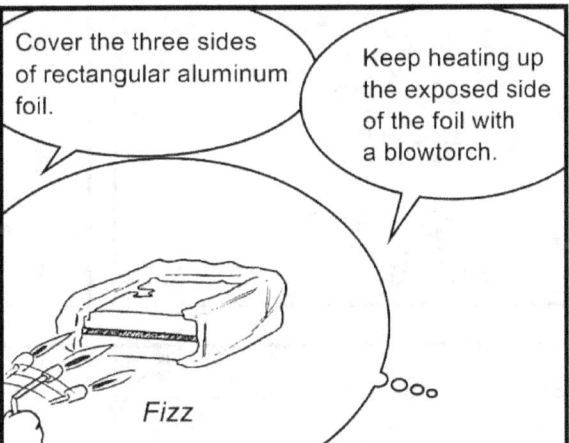

$$\frac{\partial u}{\partial t} = c^2 \left(\frac{\partial^2 u}{\partial x^2} + \frac{\partial^2 u}{\partial y^2} \right)$$

PARTIAL DIFFERENTIAL EQUATIONS

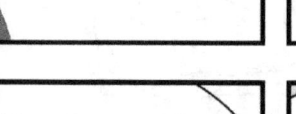

PATIAL DIFFERENTIAL EQUATIONS

The heat equation in a steady state -> Laplace's equation

What is the steady state?

The state where the temperatures at all the parts of the foil don't change any more is called the steady state.

For instance, like this

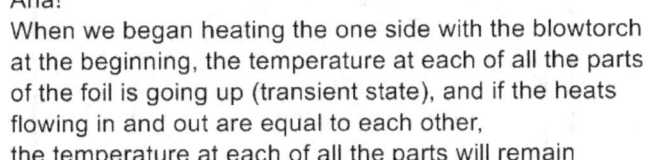

Aha!
When we began heating the one side with the blowtorch at the beginning, the temperature at each of all the parts of the foil is going up (transient state), and if the heats flowing in and out are equal to each other,
the temperature at each of all the parts will remain constant (steady state).

So, since the temperature doesn't depend on time, we get

$$\frac{\partial u}{\partial t} = 0$$

where u is temperature

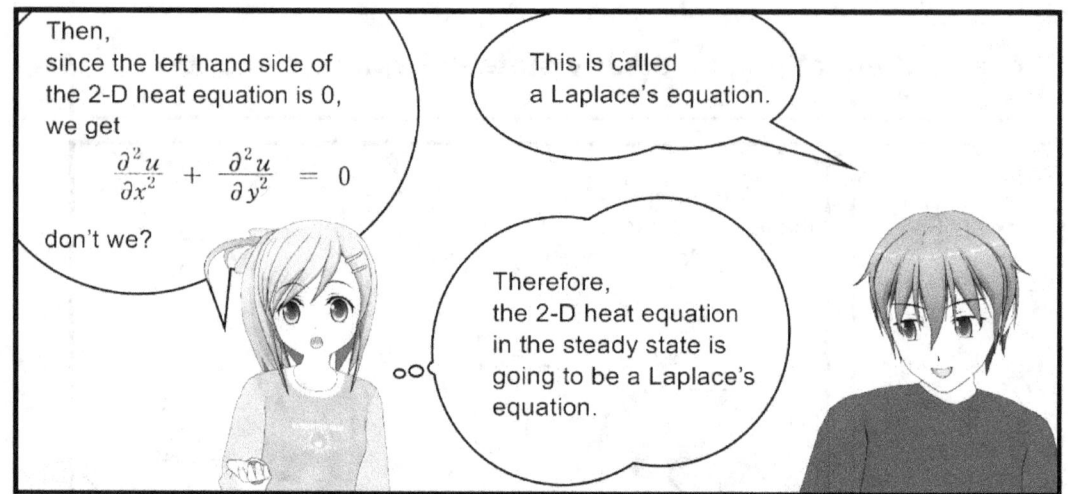

PATIAL DIFFERENTIAL EQUATIONS

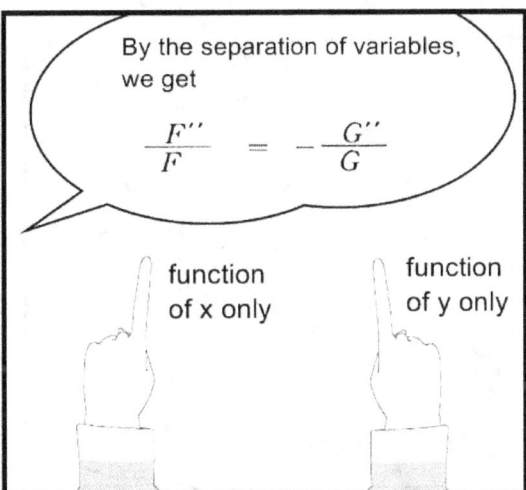

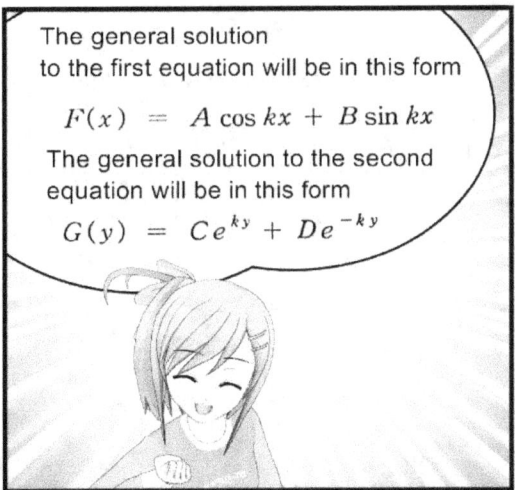

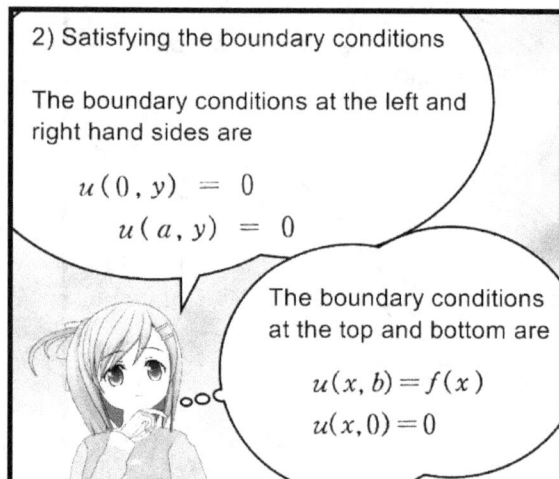

PATIAL DIFFERENTIAL EQUATIONS

Meanwhile, from the boundary conditions at the top and bottom
$u(x, b) = f(x)$
$u(x, 0) = 0$

$u(x, b) = F(x) G(b) = f(x)$
$u(x, 0) = F(x) G(0) = 0$
So, $G(0) = 0$
$u(x, b) = F(x) G(b) = f(x)$

If $F(x) = 0$
$u = 0 \rightarrow$ nonsence

Now, since
$G(y) = Ce^{ky} + De^{-ky}$
we get
$G(0) = C + D = 0$
$D = -C$

Since we have $k = \dfrac{n\pi}{a}$ put it into the equation. Then,

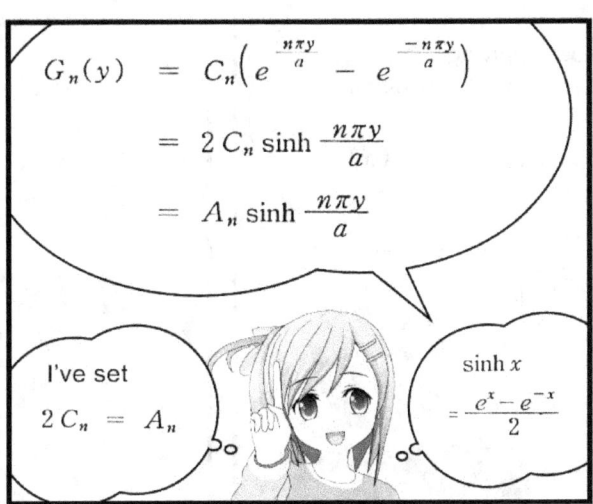

$G_n(y) = C_n\left(e^{\frac{n\pi y}{a}} - e^{-\frac{n\pi y}{a}}\right)$
$= 2C_n \sinh \dfrac{n\pi y}{a}$
$= A_n \sinh \dfrac{n\pi y}{a}$

I've set $2C_n = A_n$

$\sinh x = \dfrac{e^x - e^{-x}}{2}$

$u = F \cdot G$
$u_n(x, y) = F_n(x) G_n(y)$
$= \left(B \sin \dfrac{n\pi}{a} x\right)\left(A_n \sinh \dfrac{n\pi}{a} y\right)$
$= A_n^* \sin \dfrac{n\pi x}{a} \sinh \dfrac{n\pi y}{a}$

I've set $B \cdot A_n = A_n^*$

Therefore, these are the eigen functions.

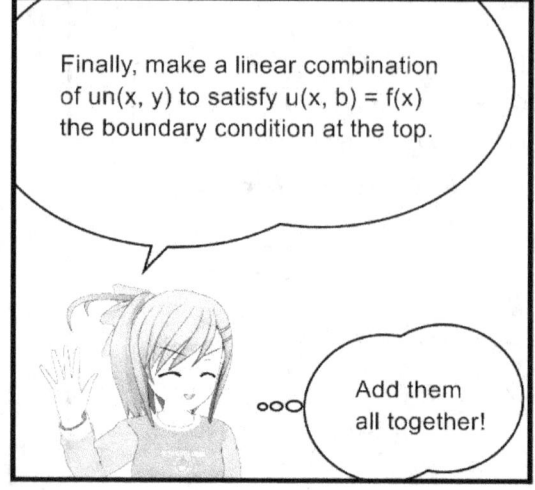

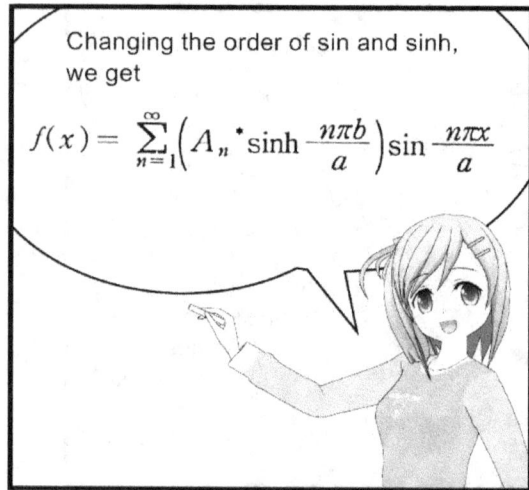

PATIAL DIFFERENTIAL EQUATIONS

That is,

$$A_n^* \sinh \frac{n\pi b}{a} = b_n = \frac{2}{a} \int_0^a f(x) \sin \frac{n\pi x}{a} dx$$

Therefore, $u(x, y) = \sum_{n=1}^{\infty} A_n^* \sin \frac{n\pi x}{a} \sinh \frac{n\pi y}{a}$

$$A_n^* \quad \left(A_n^* = \frac{2}{a \sinh(n\pi b/a)} \int_0^a f(x) \sin \frac{n\pi x}{a} dx \right)$$

Therefore, the solution satisfying the given boundary conditions of the Laplace's equation

$$\frac{\partial^2 u}{\partial x^2} + \frac{\partial^2 u}{\partial y^2} = 0$$

is

$$u(x, y) = \sum_{n=1}^{\infty} A_n^* \sin \frac{n\pi x}{a} \sinh \frac{n\pi y}{a}$$

$$\left(A_n^* = \frac{2}{a \sinh(n\pi b/a)} \int_0^a f(x) \sin \frac{n\pi x}{a} dx \right)$$

PARTIAL DIFFERENTIAL EQUATIONS

Quiz 2 Suppose the top edge of the thin glass panel is loaded with a voltage $f(x)$ as shown in the figure on the right. Then, find the voltage at each of all the parts of the panel when we earth the other thee edges and get the panel to be in the steady state.

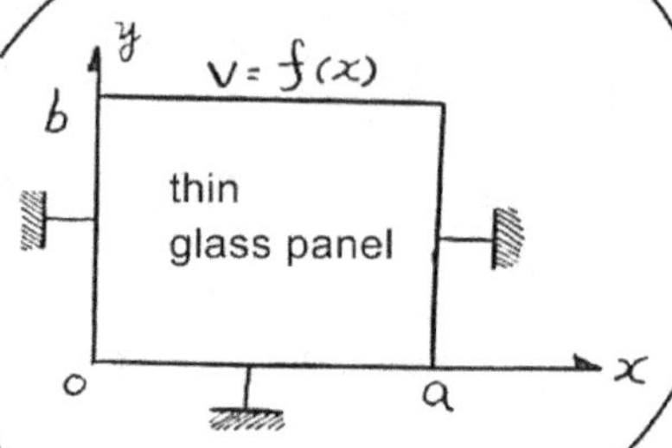

Answer

Due to the steady state,

$$\frac{\partial V}{\partial t} = 0$$

So, the voltage at each of all the parts of the panel is indicated by the 2-D Laplace's equation

$$\frac{\partial^2 V}{\partial x^2} + \frac{\partial^2 V}{\partial y^2} = 0$$

Since the boundary conditions are equivalent to those in case of the aluminum foil covered in the main text, the voltage distribution $V(x, y)$ is the same as the heat distribution $u(x, y)$ found in the main text.

Despite completely different phenomena, the same expression indicates them if we put them mathematically.

The fascination of math, The greatness of math

PATIAL DIFFERENTIAL EQUATIONS

PARTIAL DIFFERENTIAL EQUATIONS

Gravity, Electrostatic Potential -> 3-D Laplace's Equation

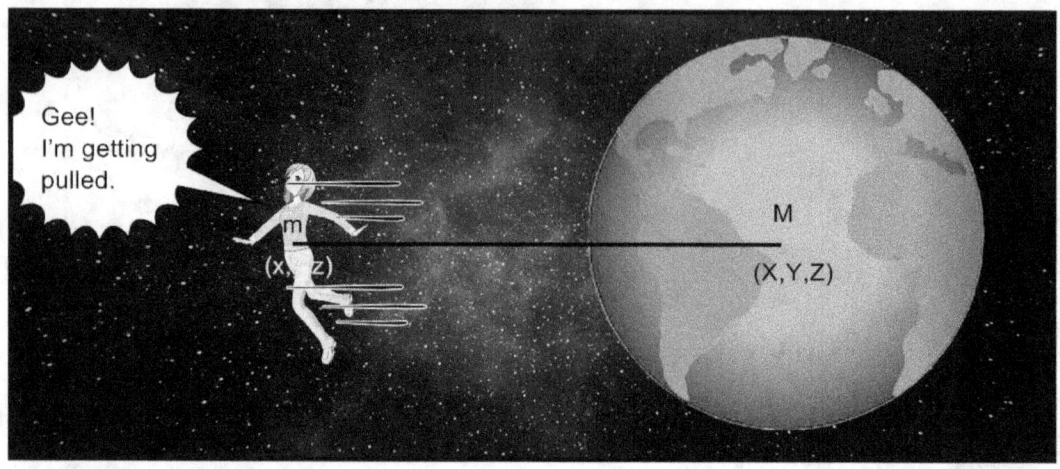

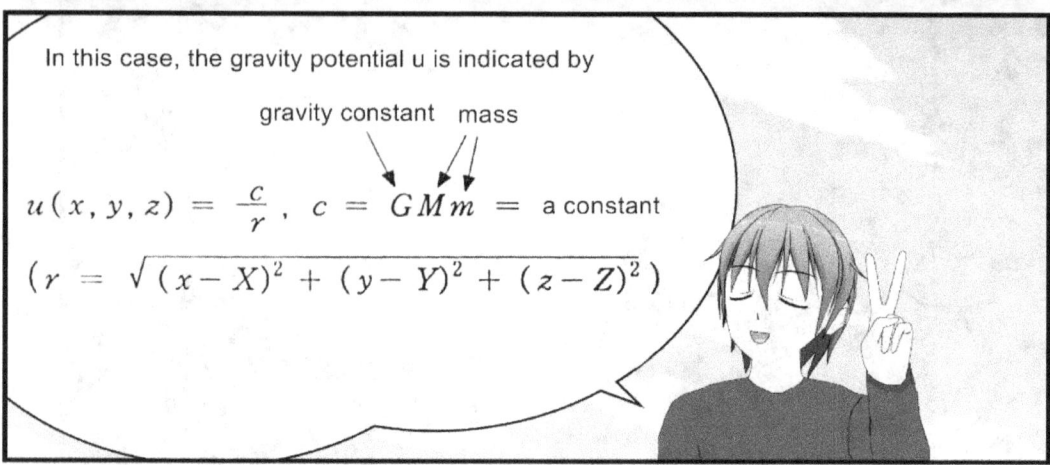

In this case, the gravity potential u is indicated by

$$u(x, y, z) = \frac{c}{r}, \quad c = GMm = \text{a constant}$$

where c involves the gravity constant G and mass M, m.

$$(r = \sqrt{(x-X)^2 + (y-Y)^2 + (z-Z)^2})$$

The gravity potential u satisfies the Laplace's equation

$$\nabla^2 u = 0$$

Really?

Well! We can readily see it if we calculate it. Refer to the quiz 1.

$$\nabla^2 u = \frac{\partial^2 u}{\partial x^2} + \frac{\partial^2 u}{\partial y^2} + \frac{\partial^2 u}{\partial z^2}$$

PATIAL DIFFERENTIAL EQUATIONS

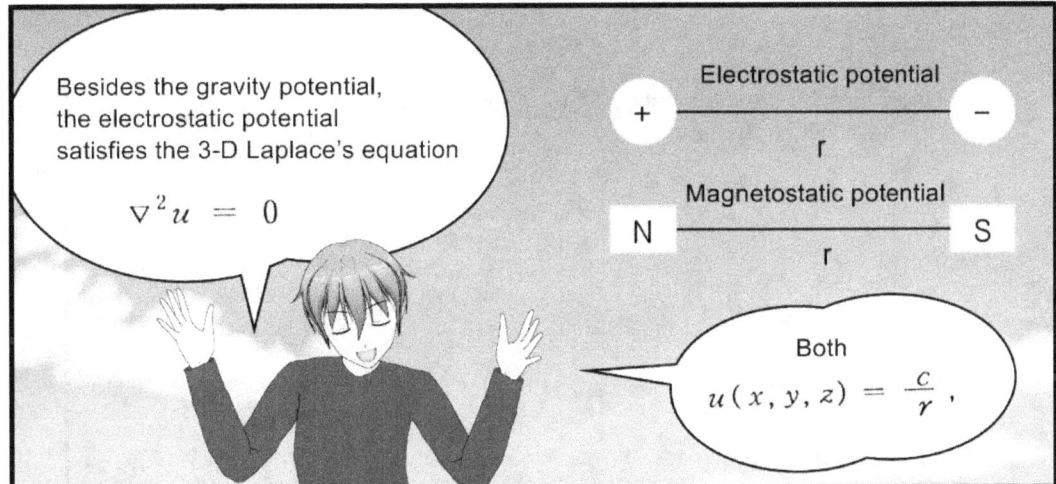

Besides the gravity potential, the electrostatic potential satisfies the 3-D Laplace's equation
$$\nabla^2 u = 0$$

Electrostatic potential

Magnetostatic potential

Both $u(x, y, z) = \dfrac{c}{r}$.

In addition, the 3-D heat conduction in a steady state will satisfy the 3-D Laplace's equation, too!

The 3-D heat equation is
$$u_t = c^2 \nabla^2 u$$
In case of a steady state, $u_t = 0$
Consequently,
$$c^2 \nabla^2 u = 0$$

For reference, putting in a polar (spherical) system the 3-D Laplace's equation, we get this!

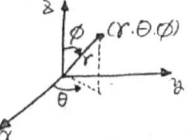

$$\nabla^2 u = \frac{1}{r^2}\left[\frac{\partial}{\partial r}\left(r^2 \frac{\partial u}{\partial r}\right) + \frac{1}{\sin\phi}\frac{\partial}{\partial\phi}\left(\sin\phi\frac{\partial u}{\partial\phi}\right) + \frac{1}{\sin^2\phi}\frac{\partial^2 u}{\partial\theta^2}\right]$$

The polar system is convenient when we have a problem where spherical symmetry is.

PARTIAL DIFFERENTIAL EQUATIONS

Quiz 1 Show that the gravitational field $u(x, y, z) = c/r$ satisfies the Laplace's equation

$$\nabla^2 u = 0$$

Answer Since

$$r = \{(x-X)^2 + (y-Y)^2 + (z-Z)^2\}^{\frac{1}{2}}$$

$$\frac{\partial u}{\partial x} = \frac{\partial}{\partial x}\left(\frac{c}{r}\right) = -\frac{2c(x-X)}{2[(x-X)^2+(y-Y)^2+(z-Z)^2]^{3/2}} = -\frac{c(x-x)}{r^3}$$

Differentiating one more time, we get

$$\frac{\partial^2 u}{\partial x^2} = \frac{\partial^2}{\partial x^2}\left(\frac{c}{r}\right) = c\left(-\frac{1}{[(x-X)^2+(y-Y)^2+(z-Z)^2]^{3/2}}\right.$$

$$\left.+\frac{3}{2}\frac{2(x-X)^2}{[(x-X)^2+(y-Y)^2+(z-Z)^2]^{5/2}}\right)$$

$$= c\left\{-\frac{1}{r^3} + \frac{3(x-X)^2}{r^5}\right\}$$

Similarly,

$$\frac{\partial^2 u}{\partial y^2} = \frac{\partial^2}{\partial y^2}\left(\frac{c}{r}\right) = c\left\{-\frac{1}{r^3} + \frac{3(y-Y)^2}{r^5}\right\}$$

$$\frac{\partial^2 u}{\partial z^2} = \frac{\partial^2}{\partial z^2}\left(\frac{c}{r}\right) = c\left\{-\frac{1}{r^3} + \frac{3(z-Z)^2}{r^5}\right\}$$

Therefore,

$$\frac{\partial^2 u}{\partial x^2} + \frac{\partial^2 u}{\partial y^2} + \frac{\partial^2 u}{\partial z^2} = c\left\{-\frac{3}{r^3} + \frac{3[(x-X)^2+(y-Y)^2+(z-Z)^2]}{r^5}\right\}$$

$$= c\left\{-\frac{3}{r^3} + \frac{3r^2}{r^5}\right\} = 0$$

Consequently,

$$\nabla^2 u = 0$$

PATIAL DIFFERENTIAL EQUATIONS

Quiz 2

Suppose an object with mass of m is in a space as shown below, find the gravity potential at all the points due to the object.

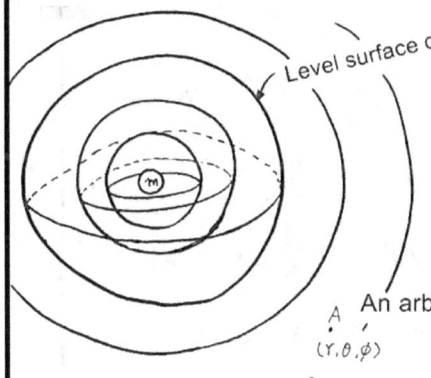

Level surface of a potential

Assume that the potential at an infinite distance is 0.

So, it is the maximum when closest.

An arbitrary point A representing the space

Answer

Since the gravity potential u is time-independent, it is in a steady state and satisfies the Laplace's equation.

Now, since the gravity potential is spherically symmetric as shown in the figure, the polar (spherical) system is convenient.

So, we need to solve the Laplace's equation in polar system

$$\nabla^2 u = \frac{1}{r^2}\left[\frac{\partial}{\partial r}\left(r^2 \frac{\partial u}{\partial r}\right) + \frac{1}{\sin\phi}\frac{\partial}{\partial \phi}\left(\sin\phi \frac{\partial u}{\partial \phi}\right) + \frac{1}{\sin^2\phi}\frac{\partial^2 u}{\partial \theta^2}\right]$$

to find the gravity potential u(x, y, z) at the arbitrary point A(r, theta, pai) representing the space.

In this case, since the gravity potential is 0 when the distance r -> infinity, the boundary condition is

$$\lim_{r \to 0} u(r, \theta, \phi) = 0$$

Answer

Separating the PDE into two ODEs first by the separation of variables to eventually separate the PDE into 3 ODEs, we get

$$u(r, \theta, \phi) = R(r) Y(\theta, \phi)$$

Putting the above into the Laplace's equation in polar system, we get

$$\frac{1}{r^2}\left[\frac{\partial}{\partial r}\left(r^2 \frac{\partial RY}{\partial r}\right) + \frac{1}{\sin\phi}\frac{\partial}{\partial \phi}\left(\sin\phi \frac{\partial RY}{\partial \phi}\right) + \frac{1}{\sin^2\phi}\frac{\partial^2 RY}{\partial \theta^2}\right] = 0$$

$$\frac{\partial}{\partial r}(r^2 R' Y) = -\left[\frac{1}{\sin\phi}\frac{\partial}{\partial \phi}(\sin\phi R Y_\phi) + \frac{1}{\sin^2\phi}(R Y_{\theta\theta})\right]$$

Dividing both sides by RY and separating the variables, we get

$$\frac{1}{R}\frac{\partial}{\partial r}(r^2 R') = -\frac{1}{Y}\left[\frac{1}{\sin\phi}\frac{\partial}{\partial \phi}(\sin\phi Y_\phi) + \frac{1}{\sin^2\phi} Y_{\theta\theta}\right]$$

The left hand side is a function r only, the right hand side is a function of theta and pai only, and so, both have to be constants. That is,

$$\frac{1}{R}\frac{\partial}{\partial r}(r^2 R') = k$$

$$-\frac{1}{Y}\left[\frac{1}{\sin\phi}\frac{\partial}{\partial \phi}(\sin\phi Y_\phi) + \frac{1}{\sin^2\phi} Y_{\theta\theta}\right] = k$$

Therefore, we get

$$\begin{cases} \frac{\partial}{\partial r}(r^2 R') - kR = 0 \rightarrow r^2 R'' + 2rR' - kR = 0 \\ \frac{1}{\sin\phi}\frac{\partial}{\partial \phi}(\sin\phi Y_\phi) + \frac{1}{\sin^2\phi} Y_{\theta\theta} + kY = 0 \end{cases}$$

PATIAL DIFFERENTIAL EQUATIONS

Answer

Next, setting

$$Y(\theta, \phi) = \Theta(\theta)\Phi(\phi)$$

and putting it into the second equation, we get

$$\frac{1}{\sin\phi}\frac{\partial}{\partial\phi}(\sin\phi\,\Theta\Phi') + \frac{1}{\sin^2\phi}\Theta''\Phi + k\Theta\Phi = 0$$

Dividing both sides by $\Theta\Phi$ yields

$$\frac{1}{\Phi}\frac{1}{\sin\phi}\frac{d}{d\phi}(\sin\phi\,\Phi') + \frac{1}{\sin^2\phi}\frac{\Theta''}{\Theta} + k = 0$$

Multiplying both sides by $\sin^2\phi$ and separating the variables, we get

$$\frac{1}{\Phi}\sin\phi\frac{d}{d\phi}(\sin\phi\,\Phi') + k\sin^2\phi = -\frac{\Theta''}{\Theta}$$

The left hand side is a function of pai only, the right hand side is a function of theta only, and so, both have to be constants. That is,

$$\frac{1}{\Phi}\sin\phi\frac{d}{d\phi}(\sin\phi\,\Phi') + k\sin^2\phi = c$$

$$-\frac{\Theta''}{\Theta} = c$$

Simplifying the above, we get

$$\sin\phi\frac{d}{d\phi}(\sin\phi\,\Phi') + (k\sin^2\phi - c)\Phi = 0$$

Dividing both sides by $\sin^2\phi$, we get

$$\begin{cases} \dfrac{1}{\sin\phi}\dfrac{d}{d\phi}(\sin\phi\,\Phi') + (k - \dfrac{c}{\sin^2\phi})\Phi = 0 \\ \Theta'' + c\Theta = 0 \end{cases}$$

Answer

Therefore, the ODEs for R, Φ, Θ are

$$\begin{cases} r^2 R'' + 2rR' - kR = 0 \\ \dfrac{1}{\sin\phi} \dfrac{d}{d\phi}(\sin\phi\, \Phi') + \left(k - \dfrac{c}{\sin^2\phi}\right)\Phi = 0 \\ \Theta'' + c\Theta = 0 \end{cases}$$

The first equation is called Euler-Cauchy equation having the solution of

$$R_n(r) = r^n \quad \text{and} \quad R_n(r) = \dfrac{1}{r^{n+1}}$$

but if it needs to satisfy the boundary condition

$$\lim_{r \to \infty} u(r, \theta, \phi) = 0$$

Rn(r) should be

$$R_n(r) = \dfrac{1}{r^{n+1}}$$

If we modify the second equation, we get the Lengendre's equation having the solution indicated by

$$\Phi_n(\phi) = P_n(\cos\phi) \quad (n = 0, 1, \ldots)$$

In this case, Pn(cos pai) is called the Legendre polynomials which are tabularized for some values of n so that we can readily read them in the table.

Since the third equation is a 2nd order ODE with simple constant coefficients, the solution is indicated in the form of

$$\Theta(\theta) = A\cos\sqrt{c}\,\theta + B\sin\sqrt{c}\,\theta$$

PATIAL DIFFERENTIAL EQUATIONS

Answer So, the gravity potential u(r, theta, pai) is indicated in the form of

$$u(r, \theta, \phi) = R\Theta\Phi$$
$$\propto \frac{1}{r^{n+1}} P_n(\cos\phi)(A\cos\sqrt{c}\,\theta + B\sin\sqrt{c}\,\theta)$$

The same is true for the electrostatic potential created when an electric charge is placed in a space.

Electrostatic potential
$$\nabla^2 V = 0$$

The same is also true for the temperature distribution in a steady state when a source of heat is placed in a space.

Temperature
$$\nabla^2 T = 0$$

4. The Solutions to PDEs by means of the Integral Transform

Laplace tansform

$$\mathcal{L}\{u(x,t)\} = \int_0^\infty e^{-st} u(x,t)\,dt$$

$$\mathcal{L}^{-1}\mathcal{L}\{u(x,t)\} = u(x,t)$$

$$\begin{pmatrix} \mathcal{L}\{u'(x,t)\} = s\mathcal{L}\{u(x,t)\} - u(x,0) \\ \mathcal{L}\{u''(x,t)\} = s^2\mathcal{L}\{u(x,t)\} - su(x,0) - u_t(x,0) \end{pmatrix}$$

Fourier Transform

$$\mathcal{F}\{u(x,t)\} = \frac{1}{\sqrt{2\pi}} \int_{-\infty}^\infty u(x,t) e^{-i\omega x}\,dx$$

$$\mathcal{F}^{-1}\mathcal{F}\{u(x,t)\} = u(x,t)$$

$$\begin{pmatrix} \mathcal{F}\{u'(x,t)\} = i\omega \mathcal{F}\{u(x,t)\} \\ \mathcal{F}\{u''(x,t)\} = (i\omega)^2 \mathcal{F}\{u(x,t)\} \end{pmatrix}$$

PATIAL DIFFERENTIAL EQUATIONS

The solutions to PDEs by means of the Laplace Transform

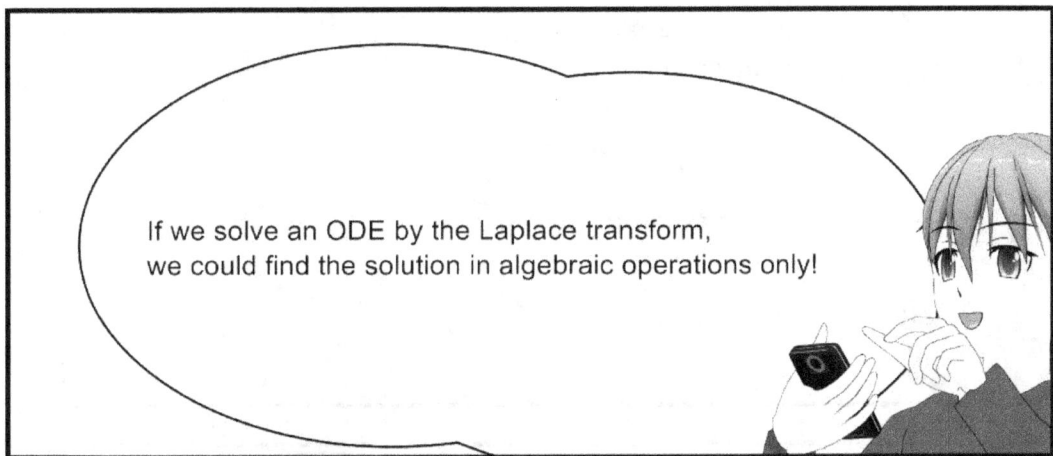

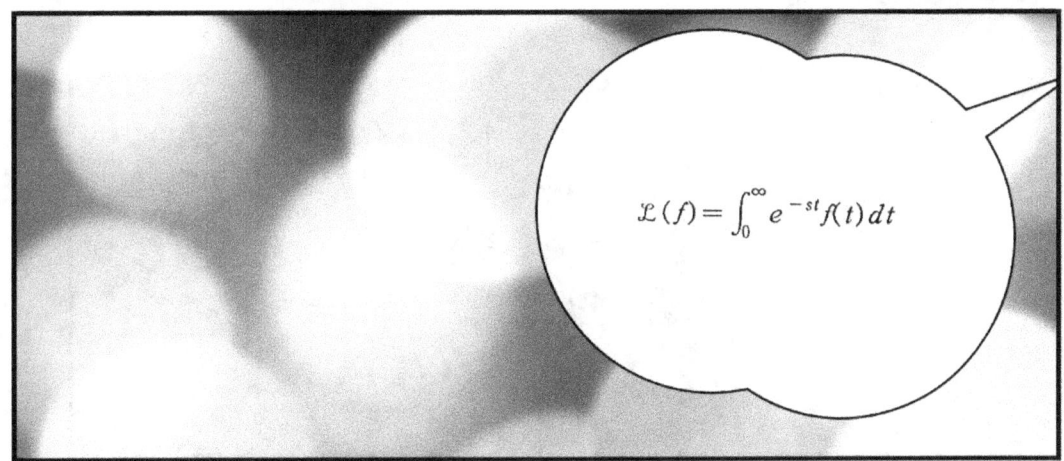

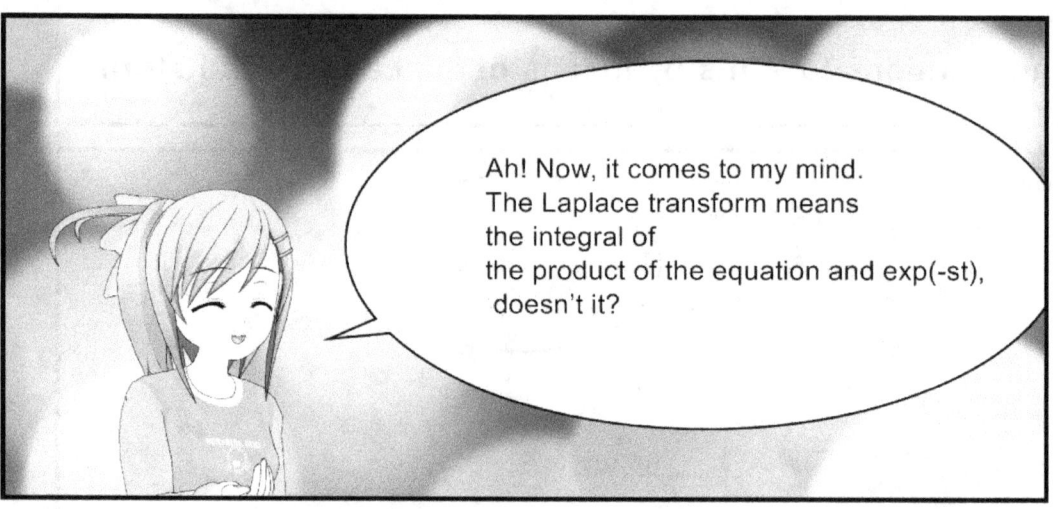

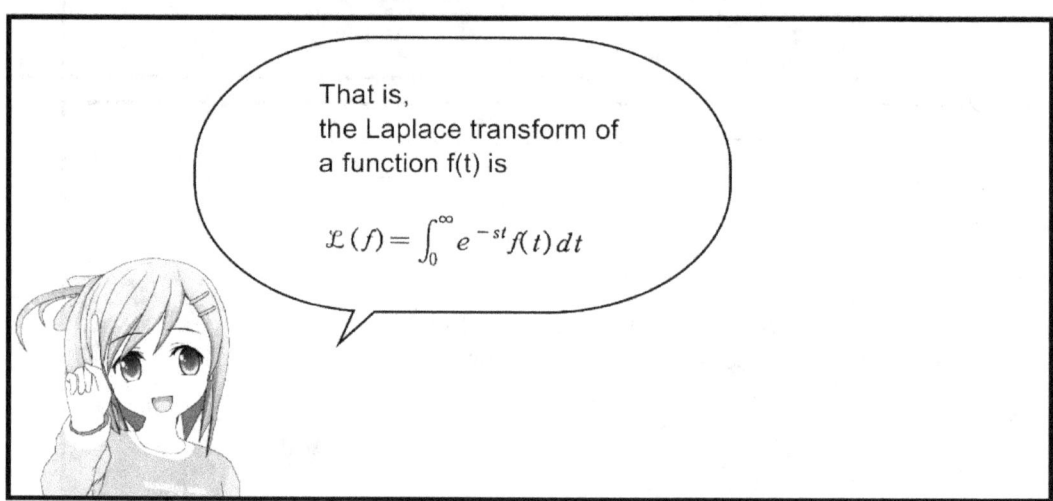

PATIAL DIFFERENTIAL EQUATIONS

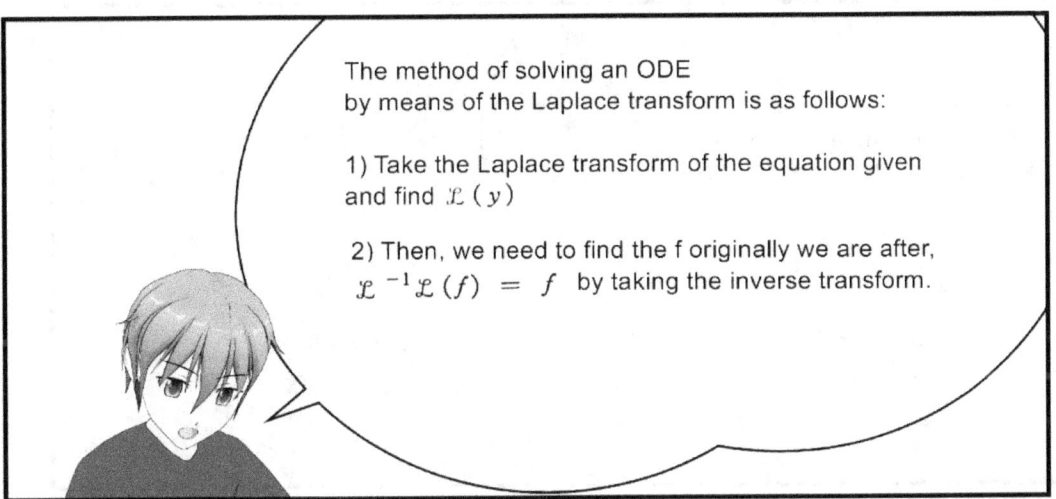

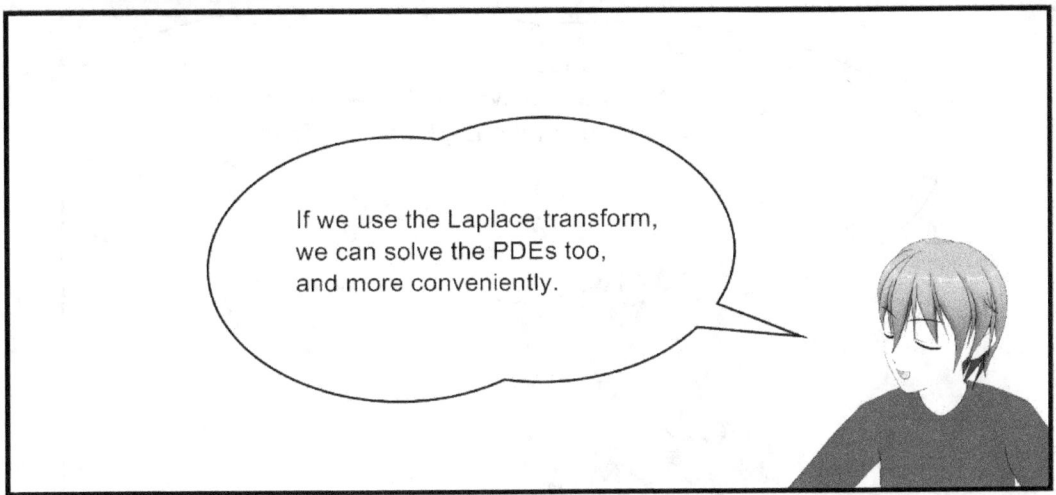

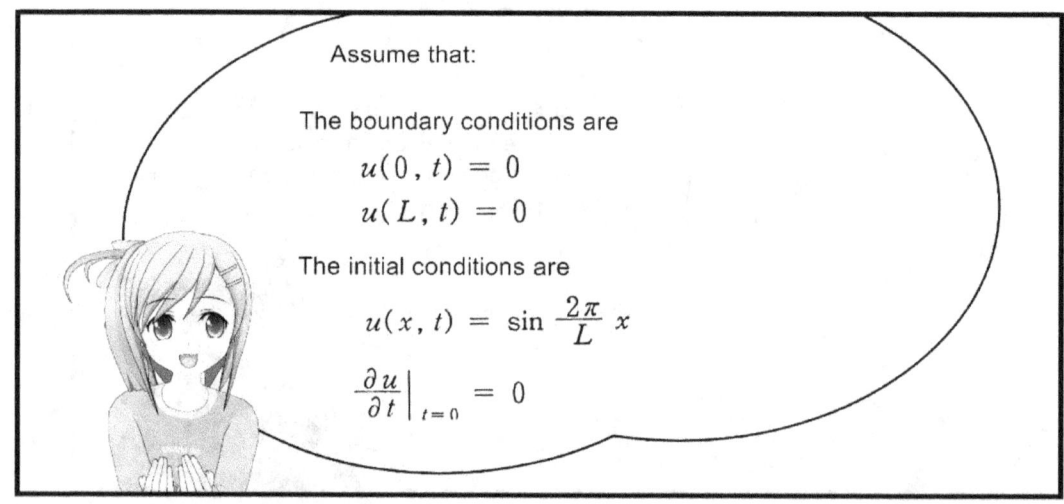

PATIAL DIFFERENTIAL EQUATIONS

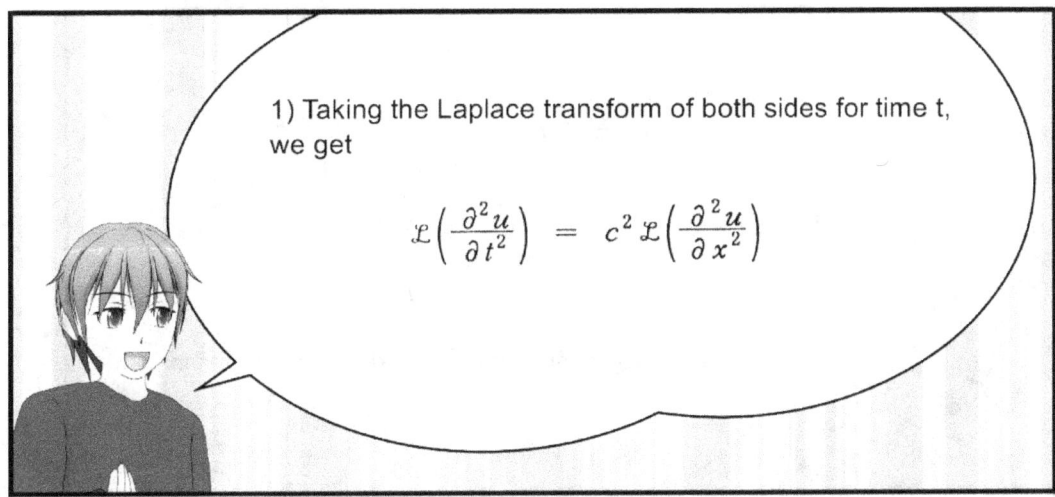

PARTIAL DIFFERENTIAL EQUATIONS

On the left hand side, we get
$$\mathcal{L}\left(\frac{\partial^2 u}{\partial t^2}\right) = s^2 \mathcal{L}(u) - su(x,0) - u_t(x,0)$$

Since
$$\mathcal{L}(f'') = s^2 \mathcal{L}(f) - sf(0) - f'(0)$$

Putting the initial conditions
$$u(x,0) = \sin\frac{2\pi}{L}x, \quad u_t(x,0) = 0$$
into the above, we get
$$\mathcal{L}\left(\frac{\partial^2 u}{\partial t^2}\right) = s^2 \mathcal{L}(u) - s \cdot \sin\frac{2\pi}{L}x$$

On the right hand side, we get
$$\begin{aligned} c^2 \mathcal{L}\left(\frac{\partial^2 u}{\partial x^2}\right) &= c^2 \int_0^\infty e^{-st} \frac{\partial^2 u}{\partial x^2} dt \\ &= c^2 \frac{\partial^2}{\partial x^2} \int_0^\infty e^{-st} u \, dt \\ &= c^2 \frac{\partial^2}{\partial x^2} \mathcal{L}(u) \end{aligned}$$

Exp(-st) doesn't include x and is taken for a constant.

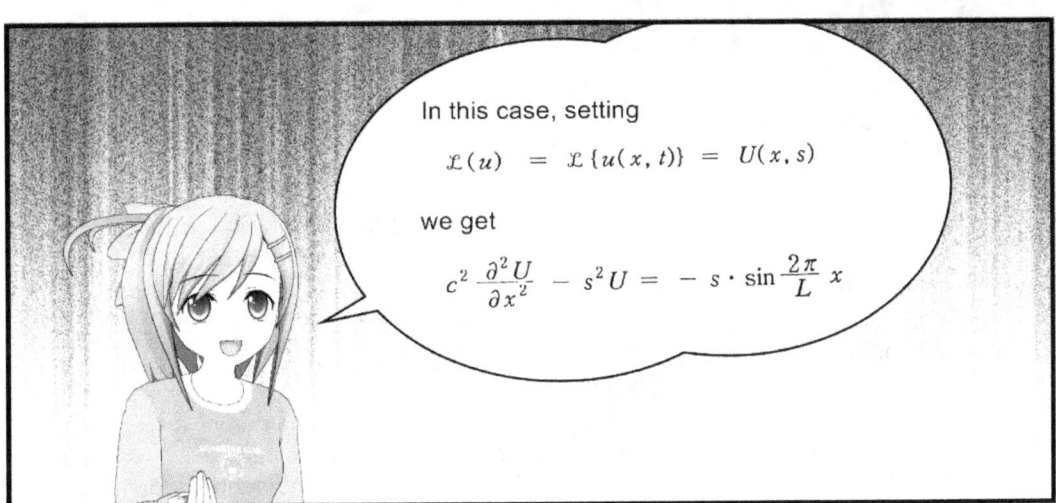

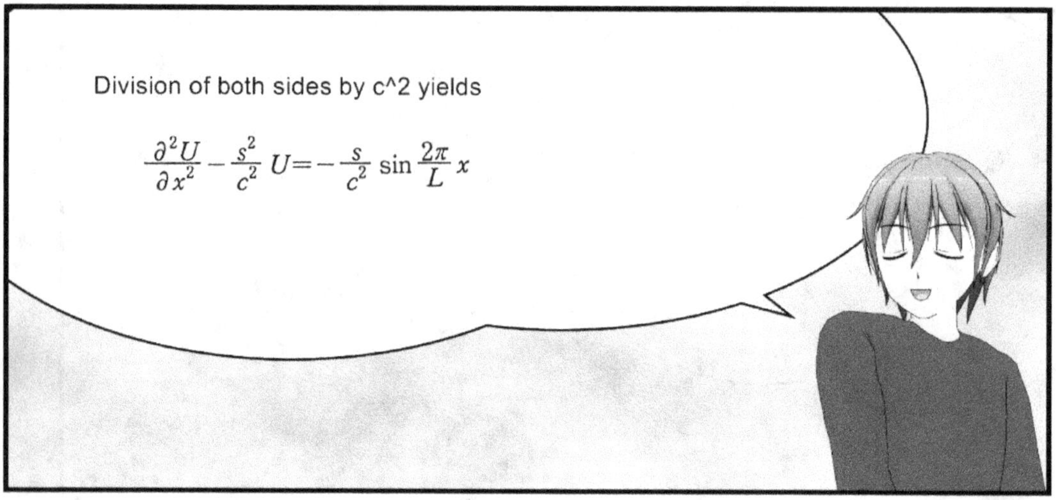

Division of both sides by c^2 yields

$$\frac{\partial^2 U}{\partial x^2} - \frac{s^2}{c^2} U = -\frac{s}{c^2} \sin \frac{2\pi}{L} x$$

Since this equation has the derivative with respect to x only, we can take it for an ODE.

It is in the form of 2nd order nonhomogeneous ODE with constant coefficients!

So, the general solution U is put in

$$U = U_h + U_p$$

the solution to the homogeneous ODE

the particular solution to the nonhomogeneous ODE

PATIAL DIFFERENTIAL EQUATIONS

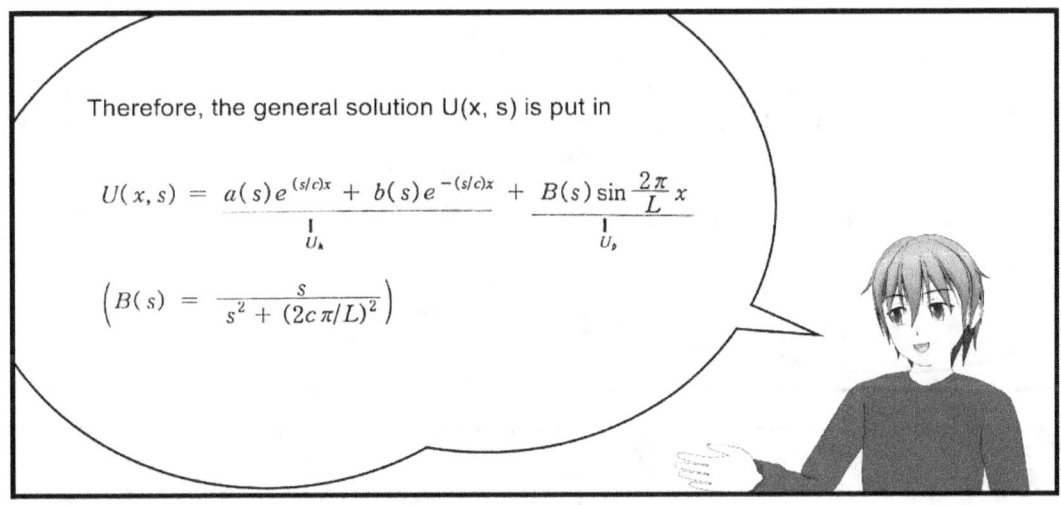

Therefore, the general solution $U(x, s)$ is put in

$$U(x, s) = \underbrace{a(s)e^{(s/c)x} + b(s)e^{-(s/c)x}}_{U_h} + \underbrace{B(s)\sin\frac{2\pi}{L}x}_{U_p}$$

$$\left(B(s) = \frac{s}{s^2 + (2c\pi/L)^2}\right)$$

We find Up by the undetermined coefficient method!

Since it is a PDE, the coefficients a, b, and B are functions of s.

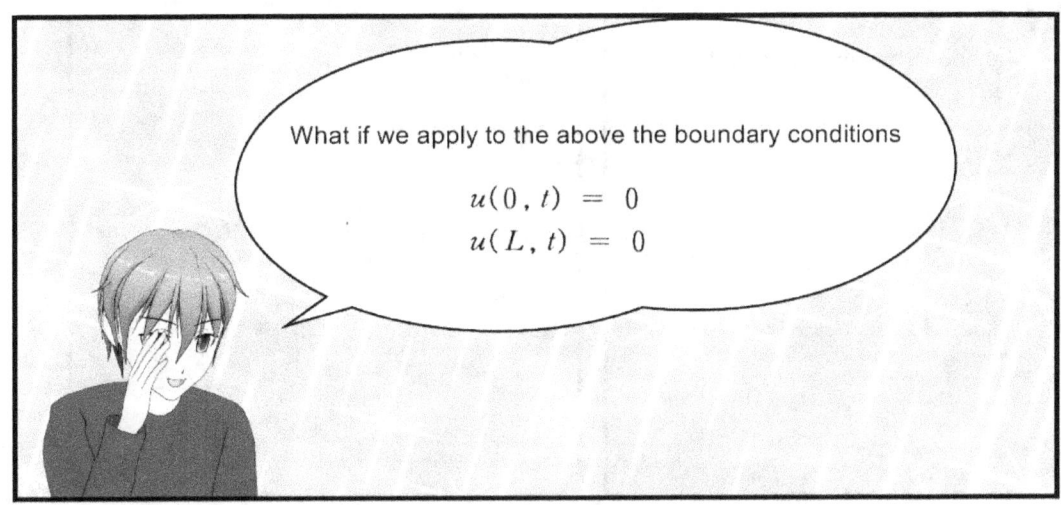

What if we apply to the above the boundary conditions

$$u(0, t) = 0$$
$$u(L, t) = 0$$

PARTIAL DIFFERENTIAL EQUATIONS

Since
$$U(x, s) = \mathcal{L}\{u(x, t)\}$$
we get
$$U(0, s) = \mathcal{L}\{u(0, t)\} = \mathcal{L}(0) = 0$$
$$U(L, s) = \mathcal{L}\{u(L, t)\} = \mathcal{L}(0) = 0$$

So, if we express the boundary condition by means of the function U, we get
$$U(0, s) = 0$$
$$U(L, s) = 0$$

Putting the above into U(x, s), we get
$$U(0, s) = a(s) + b(s) = 0$$
$$U(L, s) = a(s)e^{(s/c)L} + b(s)e^{-(s/c)L} = 0$$

If the above has to always hold, it has to be that
$$a(s) = 0, \quad b(s) = 0$$

So, putting a(s) = 0 and b(s) = 0 into the U, we get
$$U(x, s) = B(s) \sin \frac{2\pi}{L} x$$
$$= \frac{s}{s^2 + (2c\pi/L)^2} \sin \frac{2\pi}{L} x$$

PATIAL DIFFERENTIAL EQUATIONS

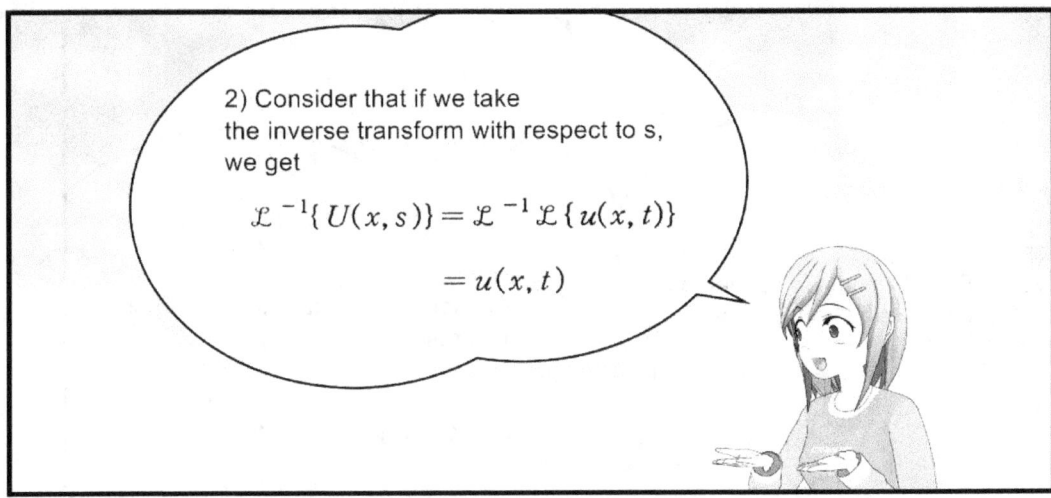

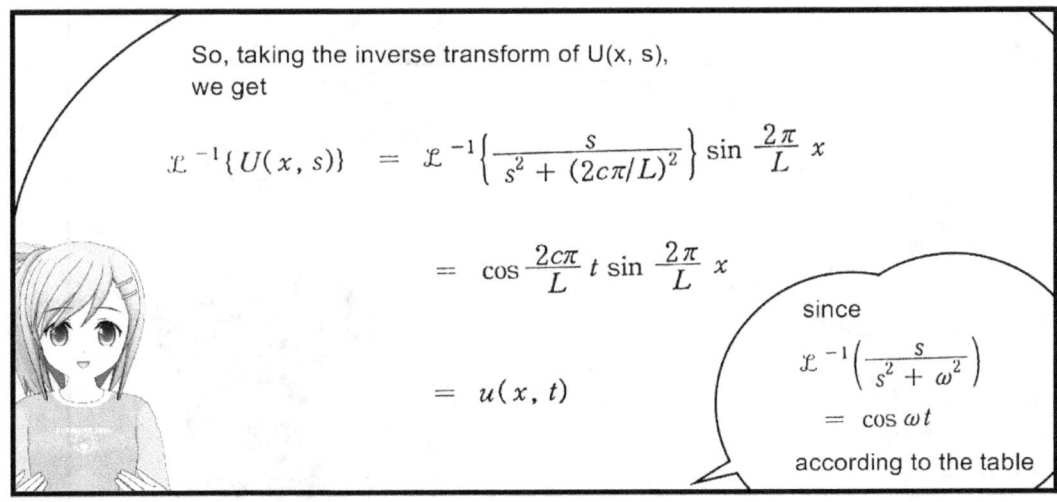

Consequently,
$$u(x, t) = \sin\frac{2\pi x}{L} \cos\frac{2c\pi t}{L}$$

Yippee!

Even if we solve it by separation of variables, we still get the same result.

After all, solving a PDE amounts to converting the PDE into a form of an ODE and solving the ODE!

Hmm...
The same is true for the method of separation of variables, isn't it?

We can solve a PDE by means of the Fourier transform, too...

Wee!

PATIAL DIFFERENTIAL EQUATIONS

Quiz 1

When we heat up the center of an iron rod to get the temperature distribution to be sin (pai/L)x, and then, insulate the lateral side of the rod and keep 0 degree C at the both ends, how does the temperature u(x, t) at each of all the parts (x) as time (t) passes away? Solve it by the Laplace transform.

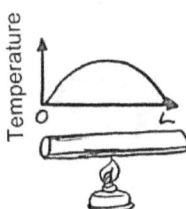

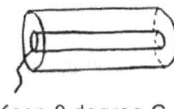

Keep 0 degree C

Answer

We need to solve the 1-D heat equation

$$\frac{\partial u}{\partial t} = c^2 \frac{\partial^2 u}{\partial x^2}$$

Note that the initial condition is

$$u(x, 0) = \sin \frac{\pi}{L} x$$

and the boundary conditions are

$$u(0, t) = 0, \quad u(L, t) = 0$$

Using the Laplace transform

1) Take the Laplace transform of both sides for time t to put them in a form of ODE.

2) Solve the ODEs.

3) Then, take the inverse transform with respect to s.

Answer

1) $\mathcal{L}\left(\dfrac{\partial u}{\partial t}\right) = c^2 \mathcal{L}\left(\dfrac{\partial^2 u}{\partial x^2}\right)$

On the left hand side, we get

$s\mathcal{L}(u) - \underline{u(x,0)}$ ($\because \mathcal{L}(u_t) = s\mathcal{L}(u) - u(x,0)$)

$\phantom{s\mathcal{L}(u) - } = \sin\dfrac{\pi}{L} x$ (the initial condition)

On the left hand side, we get

$c^2 \displaystyle\int_0^\infty e^{-st} \dfrac{\partial^2 u}{\partial x^2}\, dt = c^2 \dfrac{\partial^2}{\partial x^2}\int_0^\infty e^{-st} u\, dt = c^2 \dfrac{\partial^2}{\partial x^2} \mathcal{L}(u)$

Consequently,

$s\mathcal{L}(u) - \sin\dfrac{\pi}{L} x = c^2 \dfrac{\partial^2}{\partial x^2}\mathcal{L}(u)$

In this case, setting

$\mathcal{L}(u) = \mathcal{L}\{u(x,t)\} = U(x,s)$

we get

$\dfrac{\partial^2 U}{\partial x^2} - \dfrac{s}{c^2} U = -\sin\left(\dfrac{\pi}{L}\right) x$ ⟶ a form of an ODE

2) Since

$\dfrac{\partial^2 U}{\partial x^2} - \dfrac{s}{c^2} U = -\sin\left(\dfrac{\pi}{L}\right) x$

is in a form of a nonhomogeneous ODE with constant coefficients, the general solution is indicated by U = Uh + Up
= (the general solution to the homogeneous ODE) +
(the particular solution to the nonhomogeneous ODE).

PATIAL DIFFERENTIAL EQUATIONS

Answer

i) Finding Uh

Assuming that the base is $e^{\lambda x}$ and putting it into the homogeneous PDE

$$\frac{\partial^2 U}{\partial x^2} - \frac{s}{c^2} U = 0$$

made by setting the right hand side of the original PDE equal to 0, we get

$$\lambda^2 e^{\lambda x} - \frac{s}{c^2} e^{\lambda x} = 0$$

$$\lambda^2 - \frac{s}{c^2} = 0$$

$$\lambda = \pm \frac{\sqrt{s}}{c}$$

So, the two bases are

$$e^{(\sqrt{s}/c)x}, \quad e^{-(\sqrt{s}/c)x}$$

Multiplying each of them by the coefficients a(s) and b(s) respectively and adding the products together, we get the general solution

$$U_h = a(s) e^{(\sqrt{s}/c)x} + b(s) e^{-(\sqrt{s}/c)x}$$

(Since Uh is a function of x and s, the coefficients are functions of s.)

ii) Finding Up

Since the right hand side is $-\sin\frac{\pi}{L} x$

assuming

$$U_p = A(s) \cos\frac{\pi}{L} x + B(s) \sin\frac{\pi}{L} x$$

using the undetermined coefficient method and putting it into the original equation, we get

$$-A(s)\left(\frac{\pi}{L}\right)^2 \cos\frac{\pi}{L} x + B(s)\left(\frac{\pi}{L}\right)^2 \sin\frac{\pi}{L} x$$

$$-\frac{s}{c^2}\left\{ A(s) \cos\frac{\pi}{L} x + B(s) \sin\frac{\pi}{L} x \right\} = -\sin\frac{\pi}{L} x$$

Therefore,

$$-A(s)\left(\frac{\pi}{L}\right)^2 - \frac{s}{c^2} A(s) = 0 \rightarrow A(s) = 0$$

$$B(s)\left(\frac{\pi}{L}\right)^2 - \frac{s}{c^2} B(s) = -1 \rightarrow B(s) = \frac{-1}{(\pi/L)^2 - (s/c^2)} = \frac{c^2}{s - (\pi c/L)^2}$$

Answer

So, the general solution to the original equation is

$$U = a(s)e^{(\sqrt{s/c})x} + b(s)e^{-(\sqrt{s/c})x} + B(s)\sin\left(\frac{\pi}{L}\right)x$$

where

$$B(s) = \frac{c^2}{s - (\pi c/L)^2}$$

To find the coefficients a(s) and b(s) by means of the boundary conditions, converting u(0, t) = 0 and u(L, t) = 0 into the boundary conditions in terms of U, we get

$$U(x, s) = \mathcal{L}\{u(x, t)\}$$
$$U(0, s) = \mathcal{L}\{u(0, t)\} = \mathcal{L}(0) = 0 \quad (\because \mathcal{L}(0) = 0)$$
$$U(L, s) = \mathcal{L}\{u(L, t)\} = \mathcal{L}(0) = 0$$

Applying the above to the general solution, we get

$$U(0, s) = a(s) + b(s) = 0$$
$$U(L, s) = a(s)e^{(\sqrt{s/c})L} + b(s)e^{-(\sqrt{s/c})L} = 0$$

If the above needs to always hold, it needs to be that a(s) = 0 and b(s) = 0.

So, the general solution satisfying the initial and boundary conditions is

$$U(x, s) = B(s)\sin\frac{\pi}{L}x$$

where

$$B(s) = \frac{c^2}{s - (\pi c/L)^2}$$

PATIAL DIFFERENTIAL EQUATIONS

Answer

3) Taking the inverse transform with respect to s, we get

$$\mathcal{L}^{-1}\{U(x,s)\} = \mathcal{L}^{-1}\left\{\frac{c^2}{s-(\pi c/L)^2}\right\}\sin\frac{\pi}{L}x$$

$$= c^2 e^{(\pi c/L)^2 t}\sin(\pi/L)x$$

$(\because \mathcal{L}^{-1}\{1/(s-a)\} = e^{at})$

Take this for a constant since the transform is with respect to s.

which is the u(x, t) we are after.

That's because

$$\mathcal{L}^{-1}\{U(x,s)\} = \mathcal{L}^{-1}\mathcal{L}\{u(x,t)\} = u(x,t)$$

So,

$$u(x,t) = c^2 e^{(\pi c/L)^2 t}\sin(\pi/L)x$$

PARTIAL DIFFERENTIAL EQUATIONS

The solution to PDE by means of the Fourier Transform

Solve the 1-D heat equation

$$\frac{\partial u}{\partial t} = c^2 \frac{\partial^2 u}{\partial x^2}$$

by means of the Fourier transform.

Note that the initial temperature of the rod is

$$u(x,0) = f(x) = \begin{cases} k & (|x|<1) \\ 0 & (|x|>1) \end{cases}$$

where k ia constant.

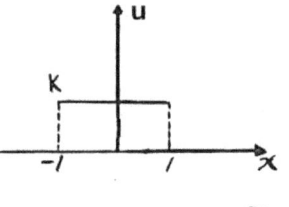

Let's assume that the boundary conditions are

$$\lim_{x \to \infty} u(x,t) = 0$$

$$\lim_{x \to \infty} u_x(x,t) = 0$$

in case where the length of the rod is infinite

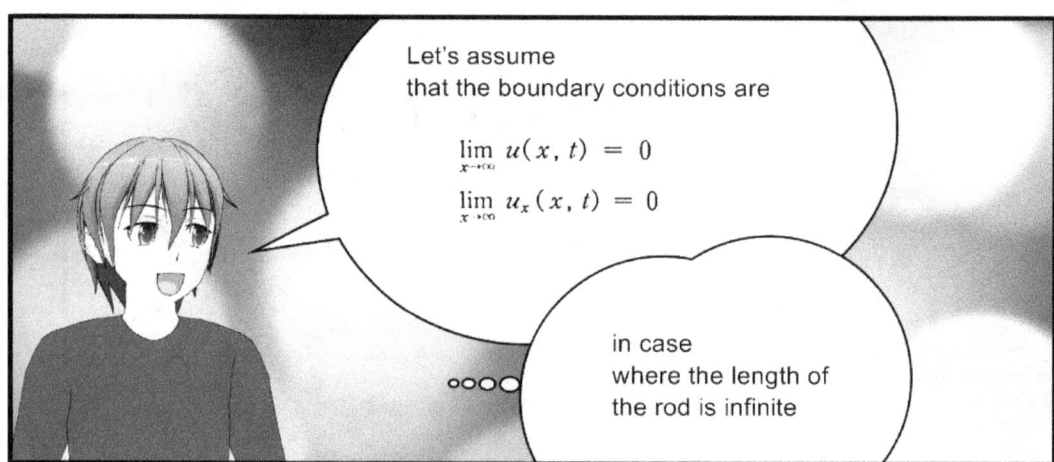

PATIAL DIFFERENTIAL EQUATIONS

The solution procedure in using the Fourier transform will be probably the same as the one in using the Laplace transform.

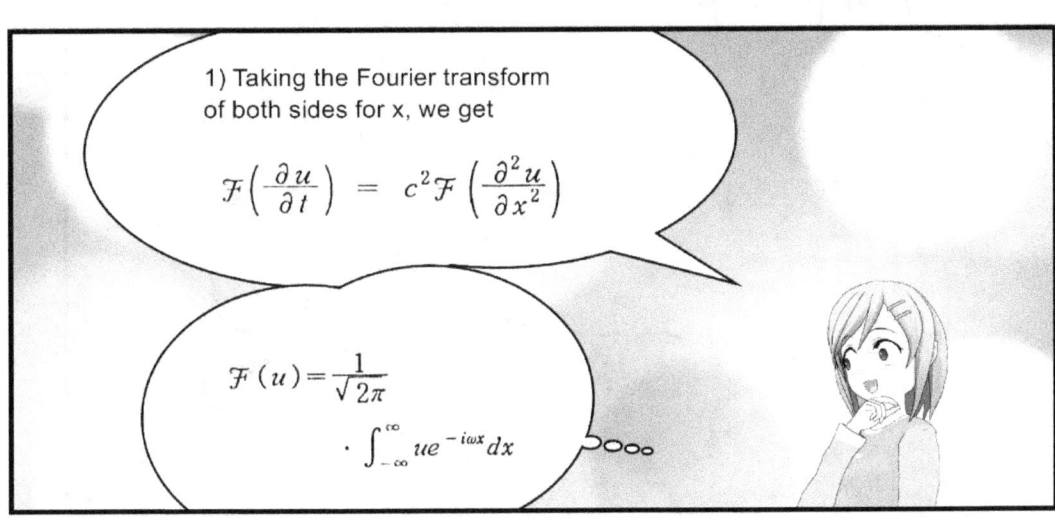

1) Taking the Fourier transform of both sides for x, we get

$$\mathcal{F}\left(\frac{\partial u}{\partial t}\right) = c^2 \mathcal{F}\left(\frac{\partial^2 u}{\partial x^2}\right)$$

$$\mathcal{F}(u) = \frac{1}{\sqrt{2\pi}} \cdot \int_{-\infty}^{\infty} u e^{-i\omega x} dx$$

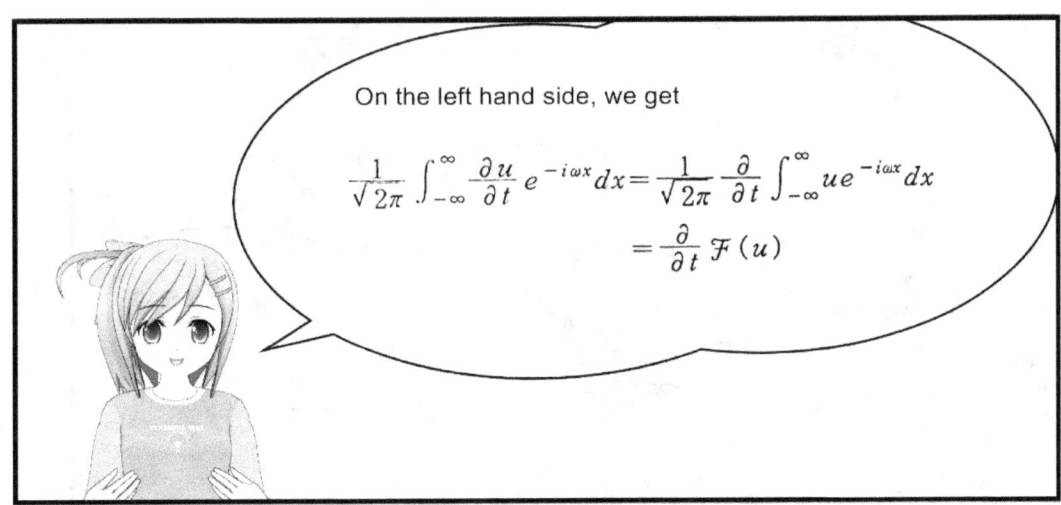

On the left hand side, we get

$$\frac{1}{\sqrt{2\pi}} \int_{-\infty}^{\infty} \frac{\partial u}{\partial t} e^{-i\omega x} dx = \frac{1}{\sqrt{2\pi}} \frac{\partial}{\partial t} \int_{-\infty}^{\infty} u e^{-i\omega x} dx$$
$$= \frac{\partial}{\partial t} \mathcal{F}(u)$$

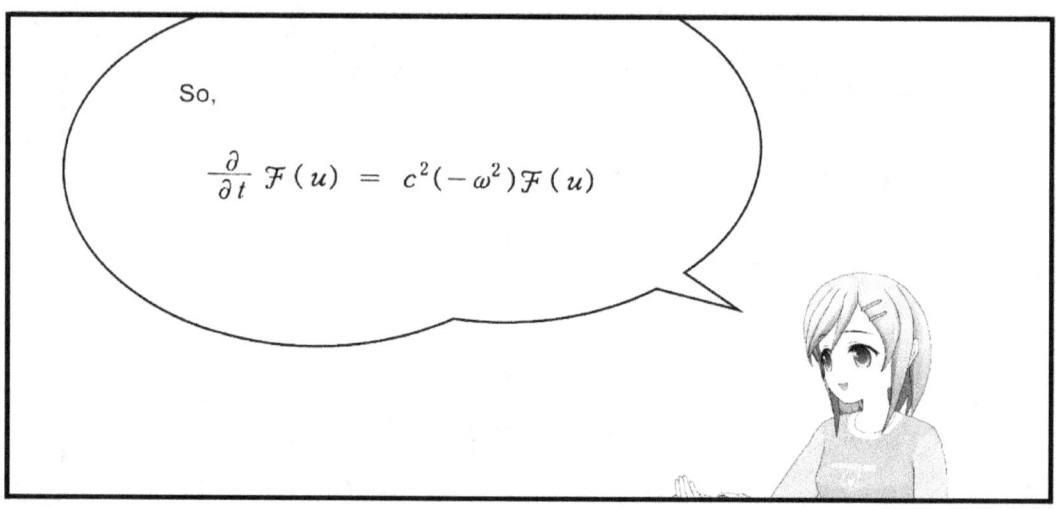
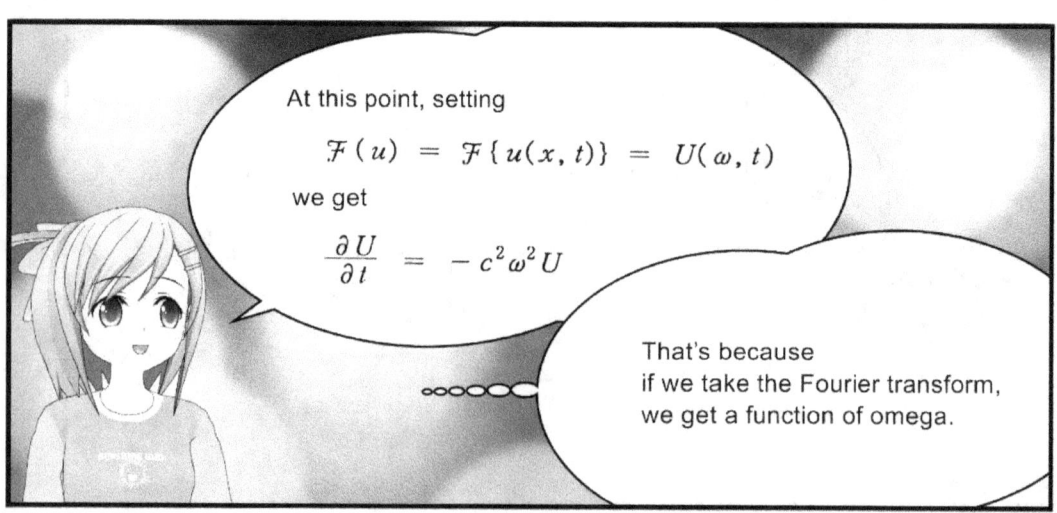

PATIAL DIFFERENTIAL EQUATIONS

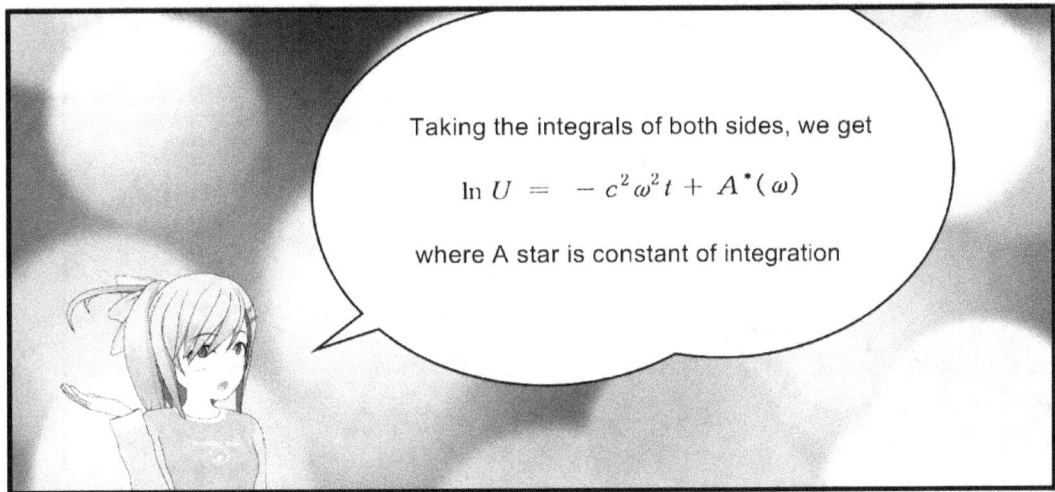

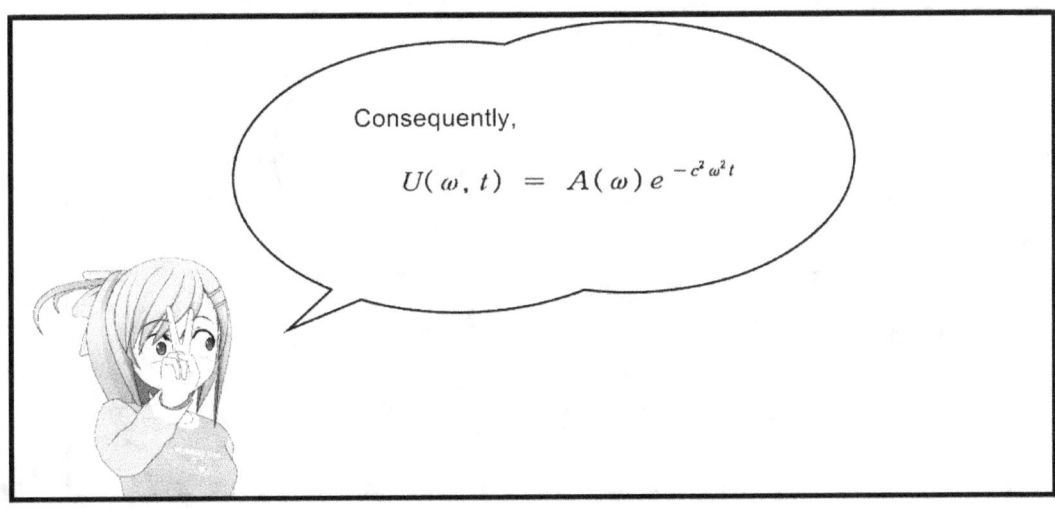

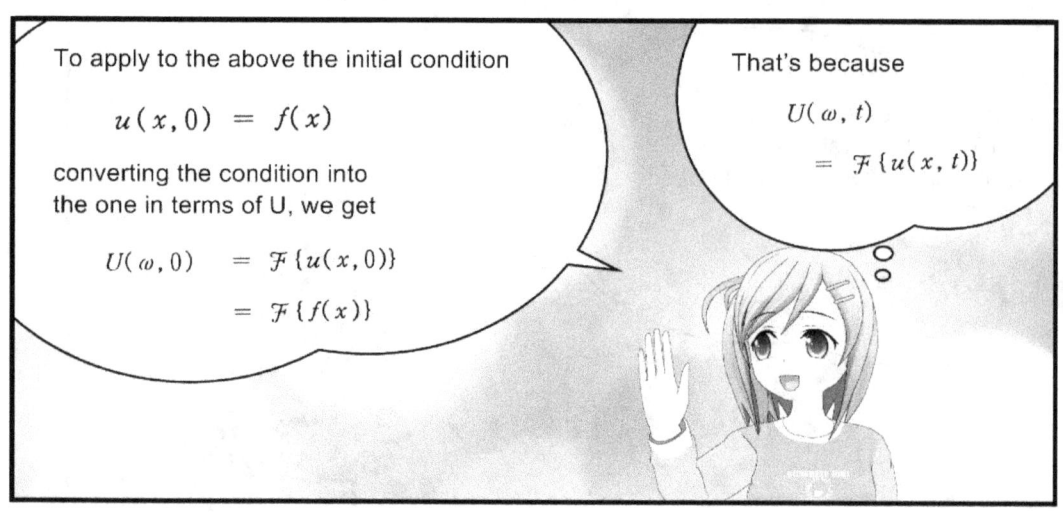

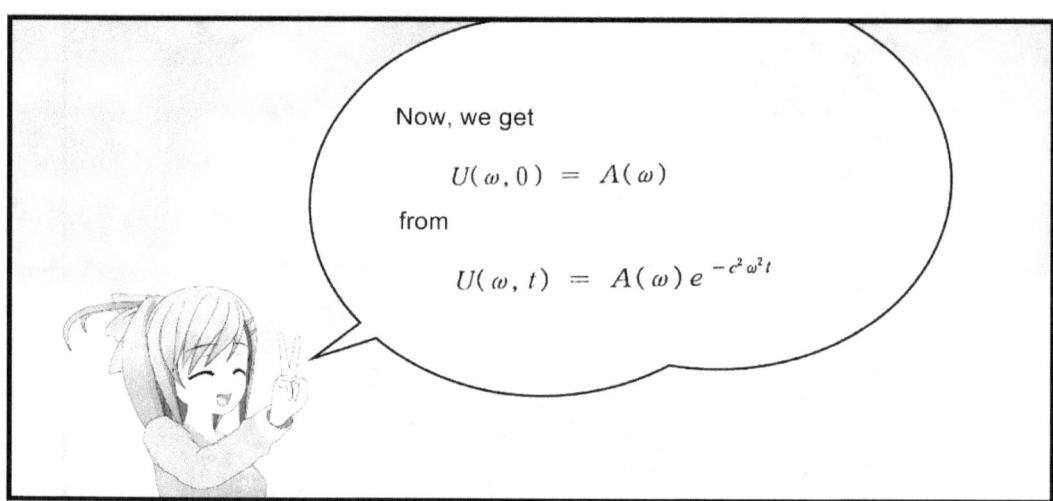

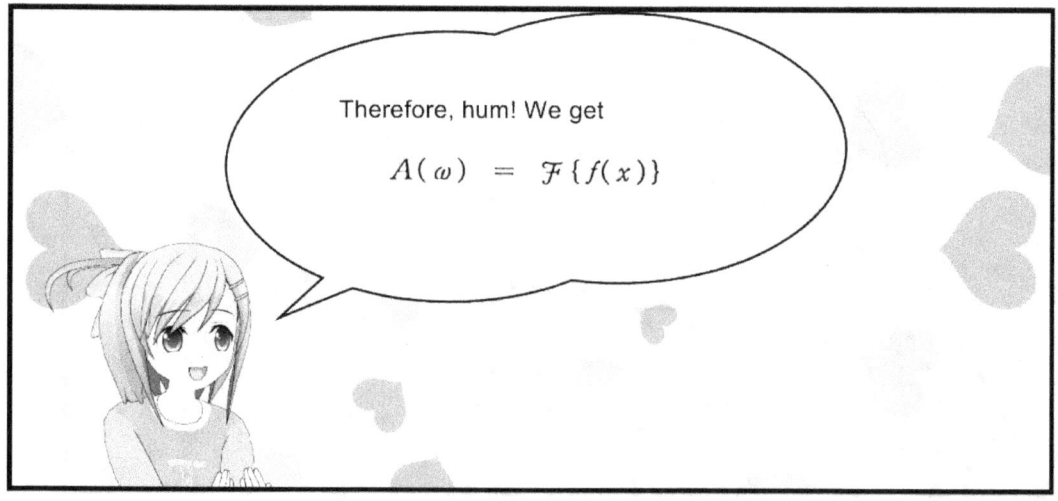

PATIAL DIFFERENTIAL EQUATIONS

Consequently, we get

$$U(\omega, t) = \mathcal{F}\{f(x)\} e^{-c^2 \omega^2 t}$$

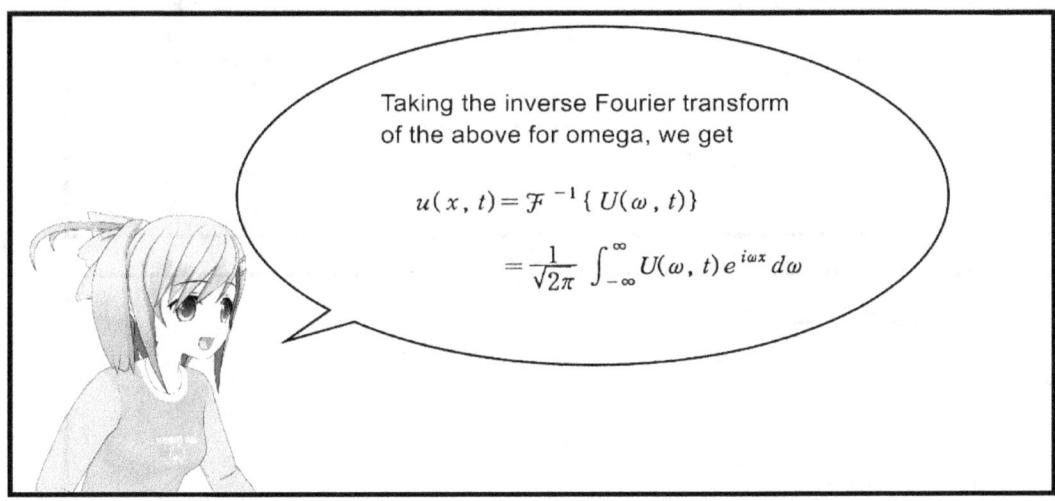

Taking the inverse Fourier transform of the above for omega, we get

$$u(x, t) = \mathcal{F}^{-1}\{U(\omega, t)\}$$
$$= \frac{1}{\sqrt{2\pi}} \int_{-\infty}^{\infty} U(\omega, t) e^{i\omega x} d\omega$$

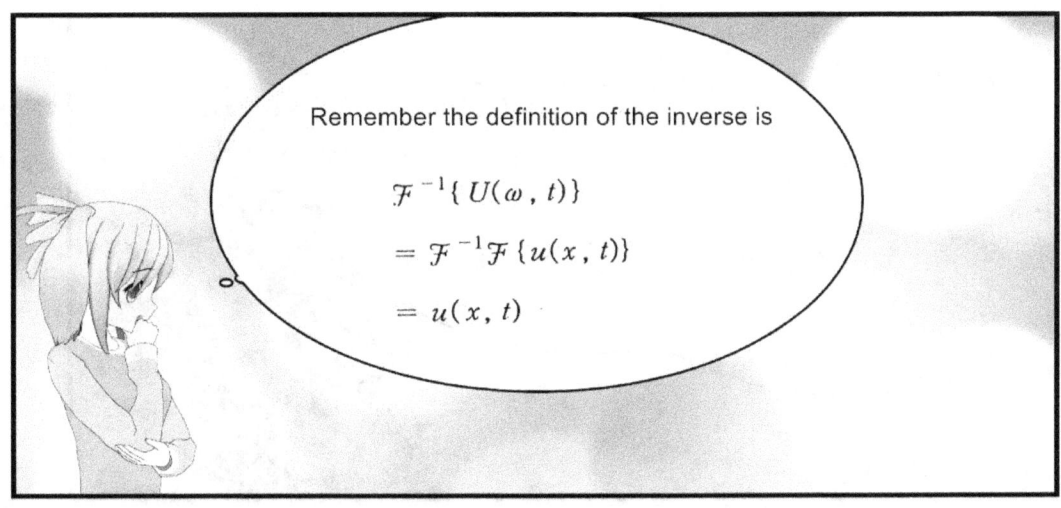

Remember the definition of the inverse is

$$\mathcal{F}^{-1}\{U(\omega, t)\}$$
$$= \mathcal{F}^{-1}\mathcal{F}\{u(x, t)\}$$
$$= u(x, t)$$

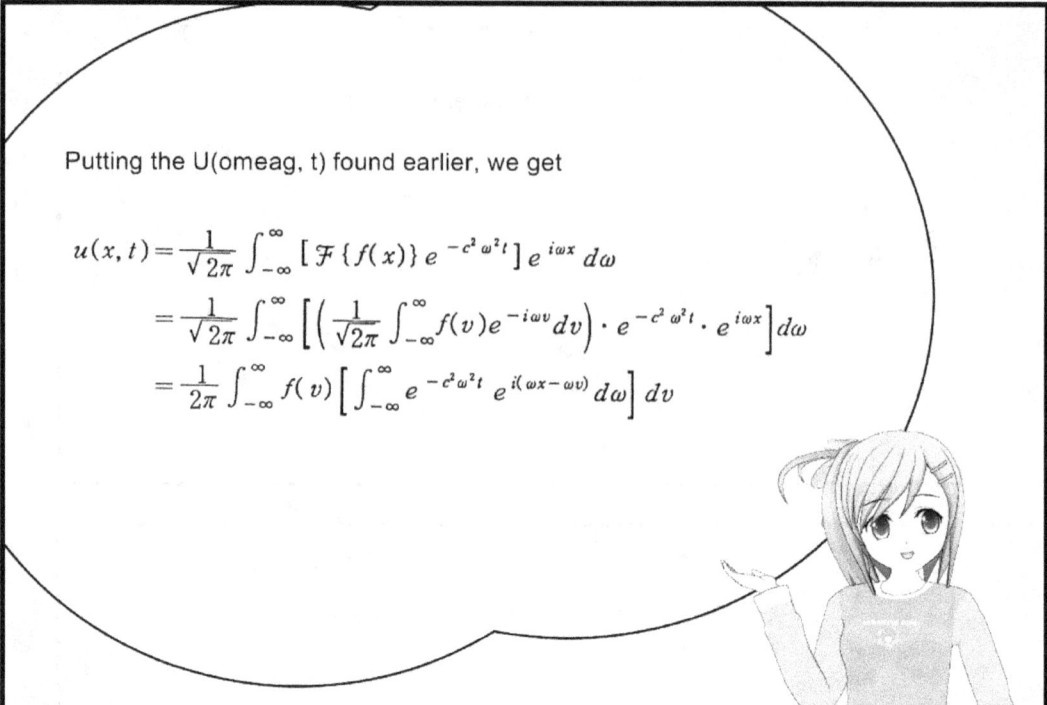

Putting the U(omeag, t) found earlier, we get

$$u(x,t) = \frac{1}{\sqrt{2\pi}} \int_{-\infty}^{\infty} [\mathcal{F}\{f(x)\} e^{-c^2\omega^2 t}] e^{i\omega x} d\omega$$

$$= \frac{1}{\sqrt{2\pi}} \int_{-\infty}^{\infty} \left[\left(\frac{1}{\sqrt{2\pi}} \int_{-\infty}^{\infty} f(v) e^{-i\omega v} dv\right) \cdot e^{-c^2\omega^2 t} \cdot e^{i\omega x}\right] d\omega$$

$$= \frac{1}{2\pi} \int_{-\infty}^{\infty} f(v) \left[\int_{-\infty}^{\infty} e^{-c^2\omega^2 t} e^{i(\omega x - \omega v)} d\omega\right] dv$$

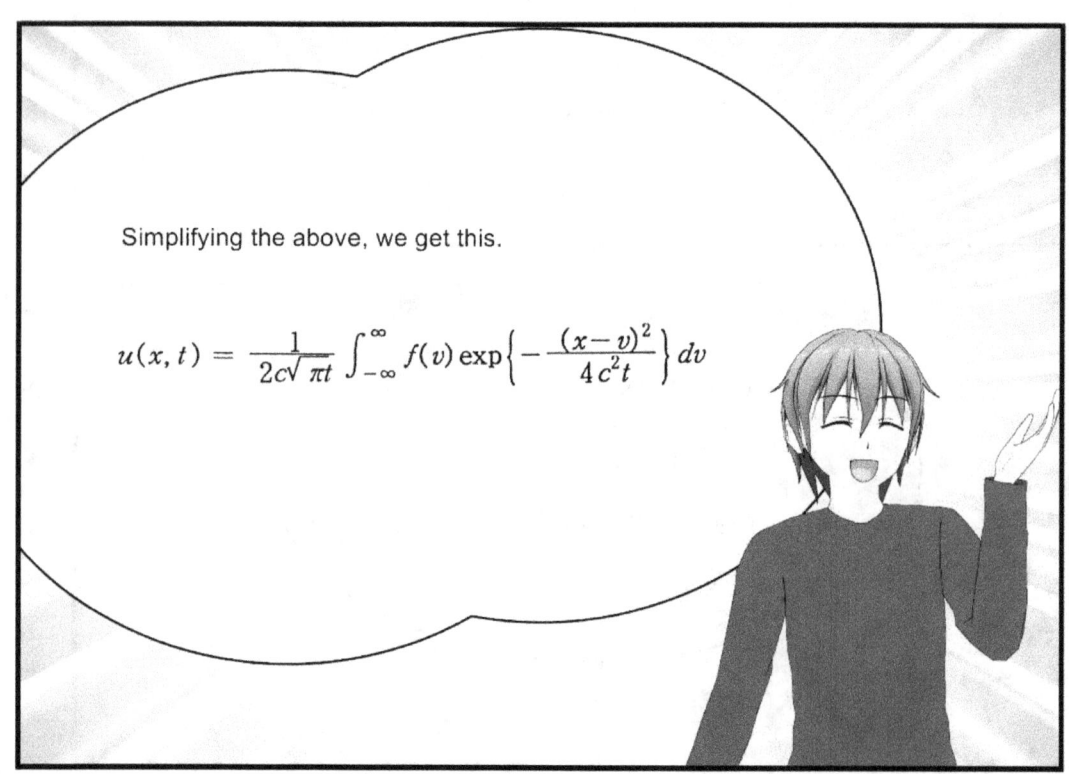

Simplifying the above, we get this.

$$u(x,t) = \frac{1}{2c\sqrt{\pi t}} \int_{-\infty}^{\infty} f(v) \exp\left\{-\frac{(x-v)^2}{4c^2 t}\right\} dv$$

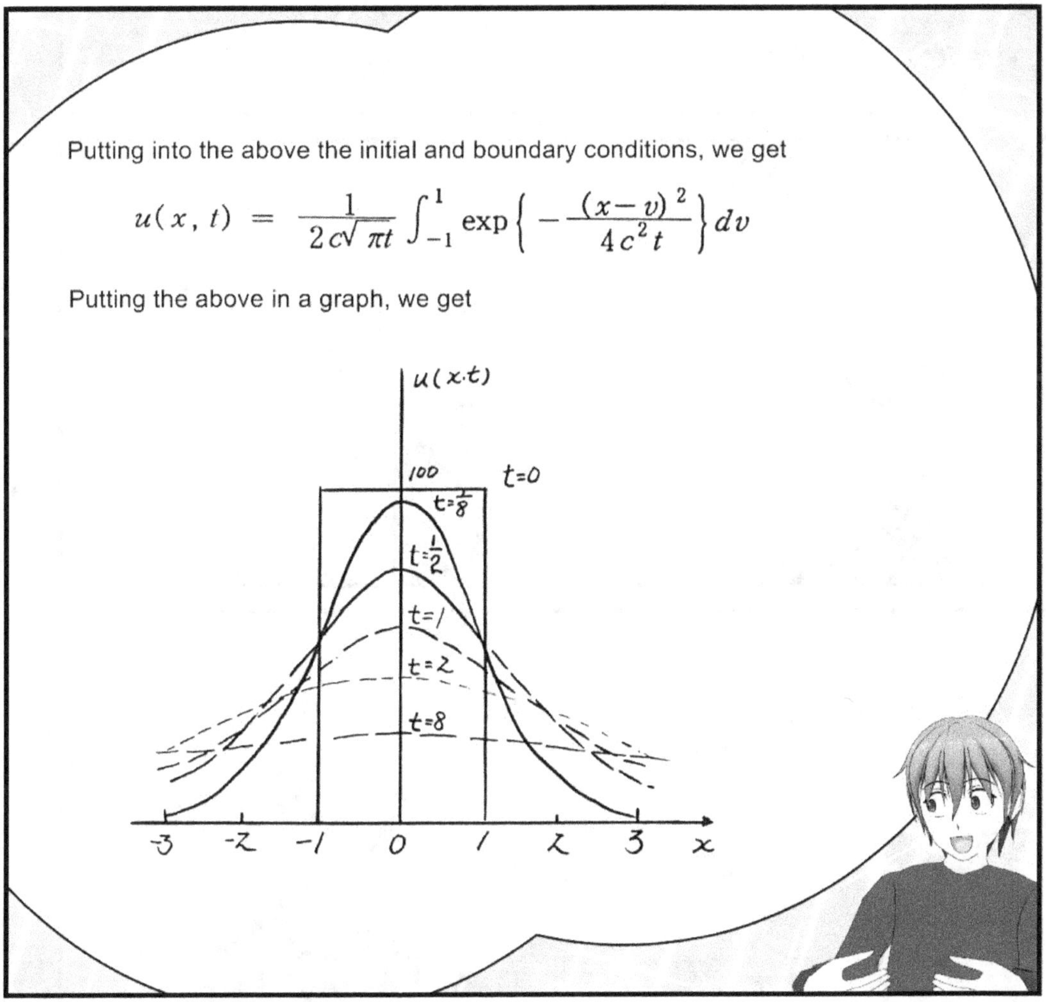

PATIAL DIFFERENTIAL EQUATIONS

Putting into the above the initial and boundary conditions, we get

$$u(x,t) = \frac{1}{2c\sqrt{\pi t}} \int_{-1}^{1} \exp\left\{-\frac{(x-v)^2}{4c^2 t}\right\} dv$$

Putting the above in a graph, we get

Professor...
Even though we use...
the Fourier transform,
it doesn't make...
any simple at all...

Ha... Ha...
Just keep in mind the fact
that we can solve a PDE
by means of the Fourier transform,
too...

The Review on PDE

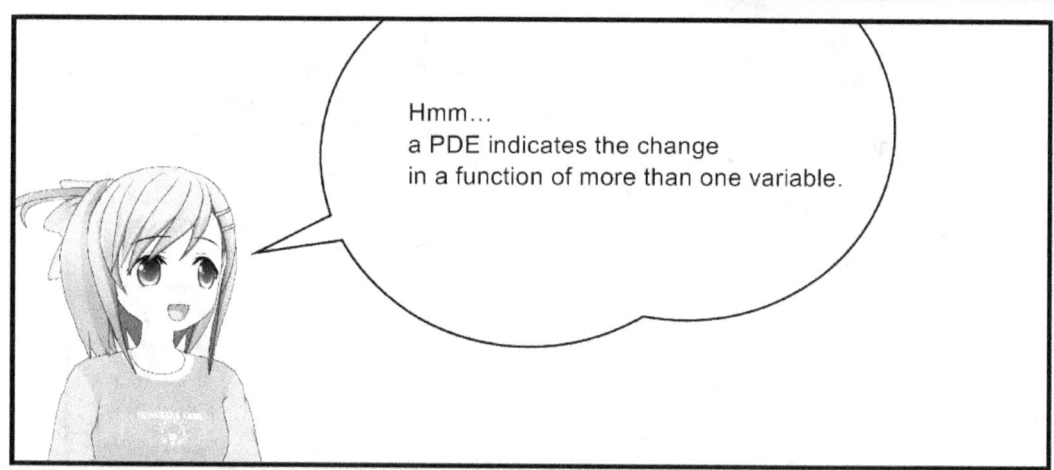

PATIAL DIFFERENTIAL EQUATIONS

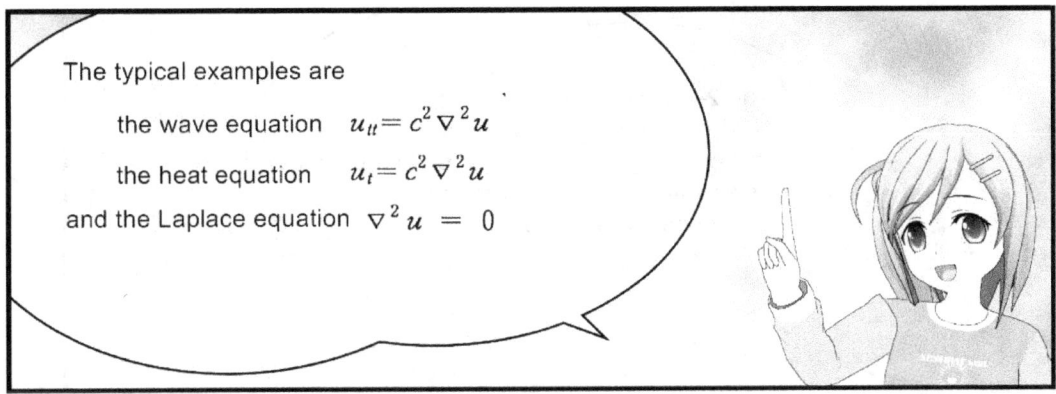

The typical examples are

the wave equation $\quad u_{tt} = c^2 \nabla^2 u$

the heat equation $\quad u_t = c^2 \nabla^2 u$

and the Laplace equation $\nabla^2 u = 0$

We solve such PDEs by means of the method of separation of variables or the integral transforms…

The 1-D wave equation indicates the vibration of a string, and is applied to such a string instrument as a guitar, violin, etc. and elastic wave, acoustic wave, etc. in a rod.

$$u_{tt} = c^2 u_{xx} \quad (c^2 = \frac{T}{\rho})$$

$$u(x, t) = \sum_{n=1}^{\infty} (B_n \cos \lambda_n t + B_n^* \sin \lambda_n t) \sin \frac{n\pi x}{L}$$

$$\begin{cases} B_n = \frac{2}{L} \int_0^L f(x) \sin \frac{n\pi x}{L} dx \\ B_n^* = \frac{2}{cn\pi} \int_0^L g(x) \sin \frac{n\pi x}{L} dx \end{cases}$$

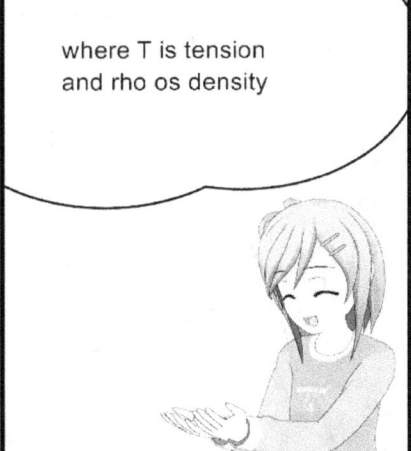

where T is tension and rho os density

PARTIAL DIFFERENTIAL EQUATIONS

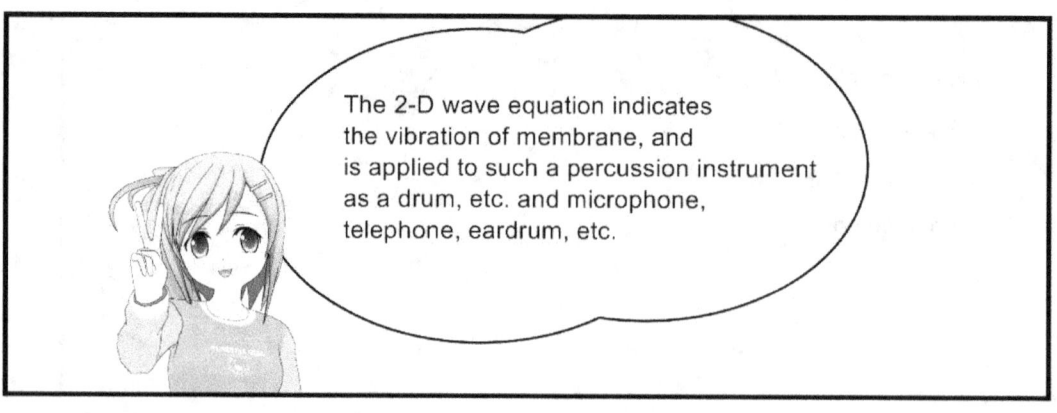

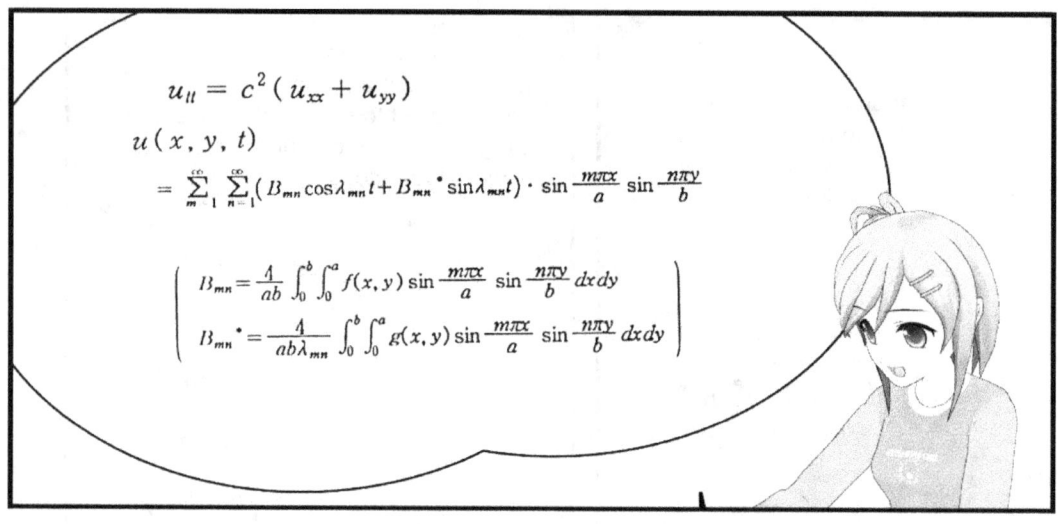

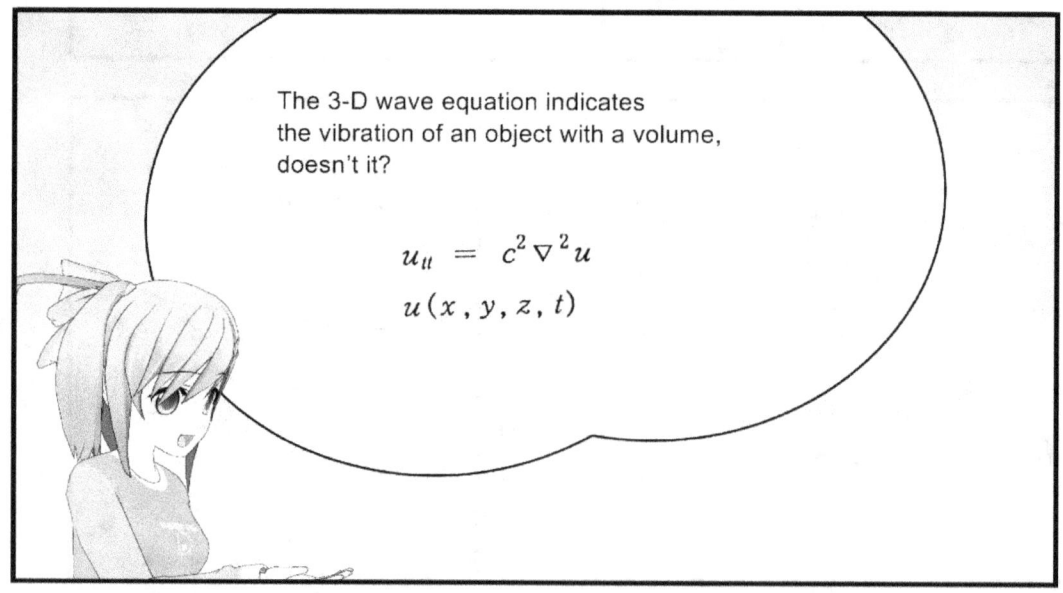

PATIAL DIFFERENTIAL EQUATIONS

The 1-D heat equation indicates heat conduction inside a rod, and the 2-D heat equation indicates heat conduction inside membrane.

$$u_t = c^2 u_{xx} \quad (c^2 = \frac{K}{\sigma \rho})$$

$$u(x, t) = \sum_{n=1}^{\infty} B_n \sin\frac{n\pi x}{L} e^{-\lambda_n^2 t}$$

$$\left(B_n = \frac{2}{L}\int_0^L f(x)\sin\frac{n\pi x}{L}\,dx\right)$$

where K is heat conductivity, sigma is specific heat, and rho is density

Of course, the 3-D heat equation

$$u_t = c^2 \nabla^2 u$$

indicates heat conduction inside an object with a volume!

In addition, the diffusion of material is governed by the same equations.

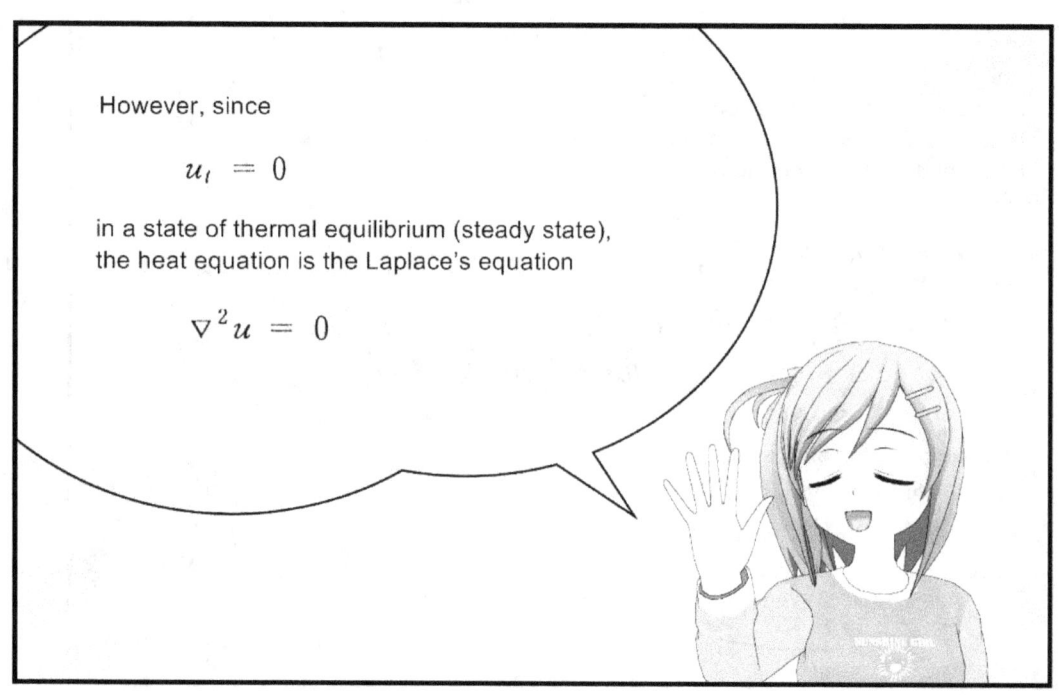

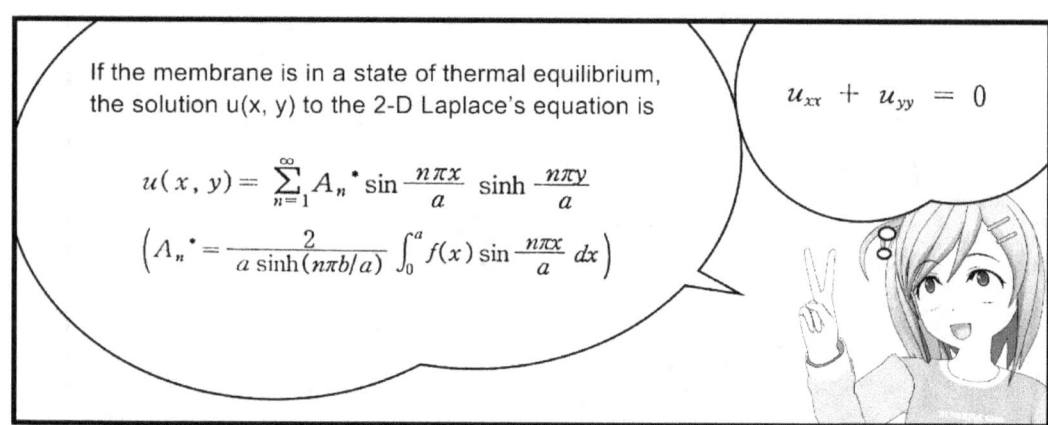

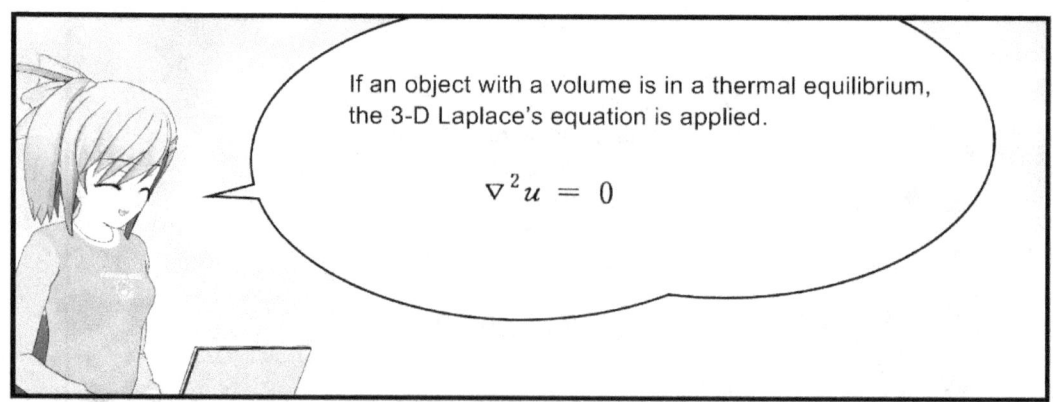

PATIAL DIFFERENTIAL EQUATIONS

In addition, the Laplace's equation is used when we indicate the gravity or electrostatic potential, too.

If the potential is spherically symmetric, we put the Laplace's equation in a polar system and get

$$\nabla^2 u = \frac{1}{r^2}\left[\frac{\partial}{\partial r}\left(r^2 \frac{\partial u}{\partial r}\right) + \frac{1}{\sin\phi}\frac{\partial}{\partial \phi}\left(\sin\phi \frac{\partial u}{\partial \phi}\right) + \frac{1}{\sin^2\phi}\frac{\partial^2 u}{\partial \theta^2}\right] = 0$$

The solution

$$u(r, \theta, \phi) \propto \frac{1}{r^{n+1}} P_n(\cos\phi) \cdot (A\cos\sqrt{c}\,\theta + B\sin\sqrt{c}\,\theta)$$

Well! You can see that a PDE is an expression implying the changes (natural laws) occurring in the world we are living (4 dimensional time-space), can't you?

Good night! See you tomorrow at school.

Bye! Professor!

PARTIAL DIFFERENTIAL EQUATIONS

www.ingramcontent.com/pod-product-compliance
Lightning Source LLC
Chambersburg PA
CBHW080547220526
45466CB00010B/3059